LONDON MATHEMATICAL SOCIETY LECTURE NOTE SERIES

Editor: PROFESSOR G. C. SHEPHARD, University of East Anglia

This series publishes the records of lectures and seminars on advanced topics in mathematics held at universities throughout the world. For the most part, these are at postgraduate level either presenting new material or describing older material in a new way. Exceptionally, topics at the undergraduate level may be published if the treatment is sufficiently original.

Prospective authors should contact the editor in the first instance.

The following titles are available

1. General cohomology theory and K-theory, PETER HILTON.
4. Algebraic topology: A student's guide, J. F. ADAMS.
5. Commutative algebra, J. T. KNIGHT.
7. Introduction to combinatory logic, J. R. HINDLEY, B. LERCHER and J. P. SELDIN.
8. Integration and harmonic analysis on compact groups, R. E. EDWARDS.
9. Elliptic functions and elliptic curves, PATRICK DU VAL.
10. Numerical ranges II, F. F. BONSALL and J. DUNCAN.
11. New developments in topology, G. SEGAL (ed.).
12. Symposium on complex analysis Canterbury, 1973, J. CLUNIE and W. K. HAYMAN (eds.).
13. Combinatorics, Proceedings of the British combinatorial conference 1973, T. P. McDONOUGH and V. C. MAVRON (eds.).
14. Analytic theory of abelian varieties, H. P. F. SWINNERTON-DYER.
15. An introduction to topological groups, P. J. HIGGINS.
16. Topics in finite groups, TERENCE M. GAGEN.
17. Differentiable germs and catastrophes, THEODOR BRÖCKER and L. LANDER.
18. A geometric approach to homology theory, S. BUONCRISTIANO, C. P. ROURKE and B. J. SANDERSON.
19. Graph theory, coding theory and block designs, P. J. CAMERON and J. H. VAN LINT.
20. Sheaf theory, B. R. TENNISON.
21. Automatic continuity of linear operators, ALLAN M. SINCLAIR.
22. Presentation of groups, D. L. JOHNSON.
23. Parallelisms of complete designs, PETER J. CAMERON.
24. The topology of Stiefel manifolds, I. M. JAMES.
25. Lie groups and compact groups, J. F. PRICE.
26. Transformation groups: Proceedings of the conference in the University of Newcastle upon Tyne, August 1976, CZES KOSNIOWSKI.
27. Skew field constructions, P. M. COHN.
28. Brownian motion, Hardy spaces and bounded mean oscillation, K. E. PETERSEN.
29. Pontryagin duality and the structure of locally compact abelian groups, SIDNEY A. MORRIS.
30. Interaction models, N. L. BIGGS.

London Mathematical Society Lecture Note Series. 31

Continuous crossed products and type III von Neumann algebras

A. VAN DAELE

Professor of Mathematics

Katholieke Universiteit te Leuven

Belgium

CAMBRIDGE UNIVERSITY PRESS

CAMBRIDGE

LONDON NEW YORK MELBOURNE

CAMBRIDGE UNIVERSITY PRESS
Cambridge, New York, Melbourne, Madrid, Cape Town, Singapore, São Paulo

Cambridge University Press
The Edinburgh Building, Cambridge CB2 8RU, UK

Published in the United States of America by Cambridge University Press, New York

www.cambridge.org
Information on this title: www.cambridge.org/9780521219754

First published 1978
Re-issued in this digitally printed version 2008

A catalogue record for this publication is available from the British Library

ISBN 978-0-521-21975-4 paperback

Dedicated to Professor L. P. Bouckaert
on the occasion of his seventieth birthday

Contents

Preface

These notes cover the material of a series of lectures given at the University of Newcastle upon Tyne on Takesaki's paper: 'Duality for crossed products and the structure of von Neumann algebras of type III' [16]. Since the appearance of Connes' thesis [2] and Takesaki's paper, the theory of crossed products has become very important in von Neumann algebras. An elementary and rather detailed treatment of the basics of this theory is given here, mainly intended for people who want an introduction to the subject. In part I, 'Crossed products of von Neumann algebras', I deal with general continuous crossed products. I introduce the notion in detail and give a proof of two important results. The first one is the commutation theorem for crossed products. It was obtained by Takesaki [16] in a special case, and by Digernes [4, 5] and Haagerup [8] in more general cases. The proof given here does not depend on the theory of dual weights, nor does it use any left Hilbert algebra. The second result given is Takesaki's duality theorem for crossed products with commutative groups.

In part II, 'The structure of type III von Neumann algebras', crossed products with modular actions are considered, that is those with the one-parameter group of *-automorphisms obtained by the Tomita-Takesaki theory, and I treat the structure theory of type III von Neumann algebras going with it [16]. Treatment is restricted to the case of σ-finite von Neumann algebras so that we can work with faithful normal states, and again our approach is different from the original one.

I would like to express my thanks to Professor J. Ringrose and to the other members of the department of pure mathematics of the University of Newcastle upon Tyne for their warm hospitality during my visit. This work was partially supported by the Science Research Council.

June 1976 A. Van Daele

Part I · Crossed products of von Neumann algebras

1. INTRODUCTION

A covariant system is a triple (M, G, α) where M is a von Neumann algebra, G is a locally compact group and α is a continuous action of G on M, that is a homomorphism $\alpha : s \rightarrow \alpha_s$ of G into the group of *-automorphisms of M such that for each $x \in M$, the map $s \rightarrow \alpha_s(x)$ is continuous from G to M where M is considered with its strong topology. To such a covariant system is associated in a natural way a new von Neumann algebra, called the crossed product of M by the action α of G, and denoted here by $M \otimes_\alpha G$ [16].

Similarly also covariant systems over C*-algebras are defined. In fact they have been known by mathematical physicists for some time already (see e.g. [7]). There they arise naturally because of time evolution of the physical system.

Here we will only be concerned with covariant systems over von Neumann algebras. Also they arise in a quite natural way, indeed the Tomita-Takesaki theory associates a strongly continuous one-parameter group of automorphisms to each faithful normal state on a von Neumann algebra [13, 15, 17], and clearly such a group is nothing else but a continuous action of $\mathbf{R}$.

The crossed product construction can be used to provide new, more complicated examples of von Neumann algebras. A special case of this, the group measure space construction, was already used by Murray and von Neumann to obtain non type I factors [6]. Recently also Connes used the crossed product construction to obtain an example of a von Neumann algebra, not anti-isomorphic to itself [3].

In connection with the Tomita-Takesaki theory however crossed products are also used to obtain structure theorems for certain types of von Neumann algebras. Among those we have the results of Connes about the structure of III_λ-factors with $0 \le \lambda < 1$ [2] and the result of Takesaki which states that every type III von Neumann algebra is isomorphic to the crossed product of a type II_∞ von Neumann algebra with some continuous action of $\mathbf{R}$ [16].

In part II of these lecture notes we treat Takesaki's structure theorem. In this part we deal with the notion of general continuous crossed products. This notion is introduced in section 2. If (M, G, α) is a covariant system, and if M acts in the Hilbert space $\mathcal{H}$, then the crossed product $M \otimes_\alpha G$ will act in the space $\mathcal{H} \otimes L_2(G)$ where $L_2(G)$ is the Hilbert space of square integrable functions on G with respect to some left Haar measure.

In section 3 we define an action θ of G on $M \otimes \mathcal{B}(L_2(G))$ where $\mathcal{B}(L_2(G))$ is the von Neumann algebra of all bounded operators on $L_2(G)$ and we show that $M \otimes_\alpha G$ can be characterized as the fixed points in $M \otimes \mathcal{B}(L_2(G))$ for the automorphisms $\{\theta_t \mid t \in G\}$. This result was obtained by Takesaki [16] in a special case, and by Digernes [4, 5] and Haagerup [8] in more general cases using the theory of dual weights. If G is compact the result can easily be obtained using the normal projection map $\int \theta_t \, dt$ onto the fixed points, where dt is the normalized Haar measure on G. In our approach we use a carefully chosen approximation procedure to make a similar argument work also if G is not compact (see also [18]). If the action α is spatial, an expression of the commutant of $M \otimes_\alpha G$ follows.

Finally in section 4 we consider the abelian case. If G is abelian it is possible to associate to a covariant system (M, G, α) a new covariant system $(\hat{M}, \hat{G}, \hat{\alpha})$ in a canonical way. For $\hat{M}$ one takes $M \otimes_\alpha G$, and $\hat{\alpha}$ is a continuous action of the dual group $\hat{G}$ on $M \otimes_\alpha G$. The action $\hat{\alpha}$ is defined in such a way that the crossed product $(M \otimes_\alpha G) \otimes_{\hat{\alpha}} \hat{G}$ of $M \otimes_\alpha G$ by the action $\hat{\alpha}$ of $\hat{G}$ is isomorphic to $M \otimes \mathcal{B}(L_2(G))$. This makes a duality structure possible if M is properly infinite and G separable so that $M \otimes \mathcal{B}(L_2(G))$ is isomorphic to M again. Then the covariant system canonically associated to $(\hat{M}, \hat{G}, \hat{\alpha})$ is in some sense equivalent to the original one (M, G, α) [16]. Our method here is very much dependent on operators similar to the unitary U in $L_2(G \times G)$ defined by $(Uf)(s, t) = f(ts, t)$. They enable us to work with tensor products and the structure becomes simpler and more transparent.

For the theory of von Neumann algebras we refer to the books of Dixmier [6] and Sakai [14]. For the Haar measure we refer to [1]. In the case where the groups are σ-compact, in particular if $G = \mathbf{R}$ as in part II, many other books treating Haar measure and abstract harmonic analysis will do [9, 11].

2. CROSSED PRODUCTS OF VON NEUMANN ALGEBRAS

Let M be a von Neumann algebra acting in a Hilbert space $\mathcal{H}$. Let G be a locally compact topological group and let $\alpha : s \to \alpha_s$ be a homomorphism of G into the group of *-automorphisms of M such that for each $x \in M$ the map $s \to \alpha_s(x)$ is continuous from G into M where M is considered with its strong topology. Such a homomorphism is called a continuous action of G on M. In the case $G = \mathbf{R}$ a continuous action is of course a strongly continuous one-parameter group of *-automorphisms. Very often the triple (M, G, α) is called a covariant system. Remark that equivalently one can consider M with the weak, σ-weak, σ-strong or σ-strong* topology. A typical example, and in fact a very important one, is obtained in the case of a von Neumann algebra M acting in $\mathcal{H}$ and a continuous unitary representation $a : s \to a_s$ of G in $\mathcal{H}$ with the property that $a_s x a_s^* \in M$ for all $x \in M$ and $s \in G$. Then $\alpha_s(x) = a_s x a_s^*$ is easily seen to define a continuous action of G on M.

To a continuous action α of a locally compact group G on a von Neumann algebra M can be associated a new von Neumann algebra, called the crossed product of M by the action α of G, and denoted by $M \otimes_\alpha G$. Notations like $R(M, \alpha)$ and $W^*(M, G)$ are also used. In this section we will carefully introduce this new von Neumann algebra. We will also give some basic properties of the operators involved. First however we will need to study the Hilbert space in which it acts.

Let ds denote a left invariant Haar measure on G and let $L_2(G)$ be the Hilbert space of (equivalence classes of) square integrable functions from G into $\mathbf{C}$ with respect to the measure ds. Then the crossed product will be a von Neumann algebra acting in the Hilbert space tensor product $\mathcal{H} \otimes L_2(G)$ of $\mathcal{H}$ and $L_2(G)$. As we shall see, in the theory of crossed products, it is important to consider elements in $\mathcal{H} \otimes L_2(G)$ as $\mathcal{H}$-valued functions on G. Let us first make this precise.

2.1 Notation. Denote by $C_c(G, \mathcal{H})$ the complex vector space of $\mathcal{H}$-valued functions on G with compact support. Let the scalar product in $\mathcal{H}$ be denoted by $\langle \cdot, \cdot \rangle$. Then for every pair $\xi, \eta \in C_c(G, \mathcal{H})$, the function $s \to \langle \xi(s), \eta(s) \rangle$ will be a continuous, complex valued function with compact support in G. Then define

$$\langle \xi, \eta \rangle = \int \langle \xi(s), \eta(s) \rangle \, ds.$$

It is then easily verified that this expression gives a scalar product on $C_c(G, \mathcal{H})$. The completion of $C_c(G, \mathcal{H})$ with respect to this scalar product is denoted by $L_2(G, \mathcal{H})$.

It is justified to call this space $L_2(G, \mathcal{H})$ as it can be shown that the set of $\mathcal{H}$-valued functions ξ on G with the following properties:

(i) $\langle \xi(\cdot), \eta_0 \rangle$ is measurable for all $\eta_0 \in \mathcal{H}$,

(ii) there is a separable subspace $\mathcal{H}_0$ such that $\xi(s) \in \mathcal{H}_0$ for all $s \in G$,

(iii) $\| \xi(\cdot) \| \in L_2(G)$

is itself a Hilbert space with scalar product defined as above, and that $C_c(G, \mathcal{H})$ is a dense subspace of this Hilbert space.

However it turns out to be much more convenient to consider elements in $L_2(G, \mathcal{H})$ as elements of the completion of $C_c(G, \mathcal{H})$ than as functions. Therefore, as we will not need this result anyway, we refer to an appendix (A) for a proof of the above statement.

In fact, what is much more important in our treatment is that $L_2(G, \mathcal{H})$ can be canonically identified with $\mathcal{H} \otimes L_2(G)$. This is done in the following proposition.

2.2 Proposition. _There is an isomorphism_ U _of_ $\mathcal{H} \otimes L_2(G)$ _onto_ $L_2(G, \mathcal{H})$ _such that_

$$(U(\xi_0 \otimes f))(s) = f(s)\xi_0$$

for any $\xi_0 \in \mathcal{H}$ and $f \in C_c(G)$, the set of complex-valued continuous functions with compact support in G.

Proof. Let $f_1, f_2, \ldots, f_n \in C_c(G)$ and $\xi_1, \xi_2, \ldots, \xi_n \in \mathcal{H}$. Define $\xi : G \to \mathcal{H}$ by $\xi(s) = \sum_{i=1}^{n} f_i(s)\xi_i$ for $s \in G$. Then clearly $\xi \in C_c(G, \mathcal{H})$ and

$$\begin{aligned}
\|\xi\|^2 = \langle \xi, \xi \rangle = \int \langle \xi(s), \xi(s) \rangle ds \\
&= \sum_{i,j=1}^{n} \int f_i(s)\overline{f_j(s)} \langle \xi_i, \xi_j \rangle ds \\
&= \sum_{i,j=1}^{n} \langle \xi_i, \xi_j \rangle \langle f_i, f_j \rangle \\
&= \langle \sum_{i=1}^{n} \xi_i \otimes f_i, \sum_{j=1}^{n} \xi_j \otimes f_j \rangle \\
&= \| \sum_{i=1}^{n} \xi_i \otimes f_i \|^2.
\end{aligned}$$

It follows that we can define a linear operator U from the algebraic tensor product of $\mathcal{H}$ with $C_c(G)$ into $C_c(G, \mathcal{H})$ by $U(\sum_{i=1}^{n} \xi_i \otimes f_i) = \xi$. Also U is isometric and therefore extends uniquely to an isometry from the Hilbert space tensor product $\mathcal{H} \otimes L_2(G)$ into $L_2(G, \mathcal{H})$, the extension is still denoted by U. It remains to show that U is onto, so that indeed U is an isomorphism. For this it is sufficient to show that functions of the form ξ above are dense in $C_c(G, \mathcal{H})$. So let $\xi_0 \in C_c(G, \mathcal{H})$, let K be the compact support of ξ_0 and let V be an open set with finite Haar measure such that $K \subseteq V$. Take $\varepsilon > 0$ and for each $s \in K$ choose a neighbourhood V_s of s such that $V_s \subseteq V$ and $\|\xi_0(t) - \xi_0(s)\| < \varepsilon$ for all $t \in V_s$. Then choose points $s_1, s_2 \ldots s_n$ in K such that $K \subseteq V_{s_1} \cup V_{s_2} \ldots \cup V_{s_n}$ and positive functions $h_1, h_2 \ldots h_n$ in $C_c(G)$ such that the support of h_j lies in V_{s_j} for all $j = 1, n$, and such that

$$0 \leq \sum_{j=1}^{n} h_j(s) \leq 1 \text{ for all } s \in G \text{ and } \sum_{j=1}^{n} h_j(s) = 1$$

for all $s \in K$. Then if we define $\xi = \sum_{j=1}^{n} h_j \xi_0(s_j)$ we get

$$\begin{aligned} \| \xi(s) - \xi_0(s) \| &= \| \sum_{j=1}^{n} h_j(s) \xi_0(s_j) - \sum_{j=1}^{n} h_j(s) \xi_0(s) \| \\ &\leq \sum_{j=1}^{n} h_j(s) \| \xi_0(s_j) - \xi_0(s) \| \\ &\leq \sum_{j=1}^{n} h_j(s) \varepsilon \leq \varepsilon \end{aligned}$$

for all $s \in G$. Finally since ξ and ξ_0 have support in the set V we get

$$\| \xi - \xi_0 \| \leq \varepsilon p^{\frac{1}{2}} \text{ where } p \text{ is the Haar measure of } V.$$

This proves the result.

In what follows we will always identify $L_2(G, \mathcal{H})$ and $\mathcal{H} \otimes L_2(G)$ by means of this isomorphism. So for any $\xi \in \mathcal{H}$ and $f \in C_c(G)$ we will consider $\xi \otimes f$ as a function on G with values in $\mathcal{H}$ given by $(\xi \otimes f)(s) = f(s)\xi$.

So far we have considered the space in which the crossed product is going to act. Now we come to the operators that will generate the crossed product.

2.3 Lemma. _For every_ $x \in M$ _and_ $\xi \in C_c(G, \mathcal{H})$ _we have that the function_ ξ_1, _defined by_ $\xi_1(s) = \alpha_{s^{-1}}(x)\xi(s)$ _is again in_ $C_c(G, \mathcal{H})$. _Moreover_ $\| \xi_1 \| \leq \|x\| \, \|\xi\|$.

Proof. Remark first that indeed ξ_1 is a function of G in $\mathcal{H}$ as $\xi(s) \in \mathcal{H}$ for every s and $\alpha_{s^{-1}}(x) \in M$ and M acts in $\mathcal{H}$. Clearly ξ_1 has also compact support, so we must show continuity. This follows from the calculation below with s_0 fixed in G and $s \to s_0$.

Indeed
$$\begin{aligned} \| \xi_1(s) - \xi_1(s_0) \| &= \| \alpha_{s^{-1}}(x)\xi(s) - \alpha_{s_0^{-1}}(x)\xi(s_0) \| \\ &\leq \| \alpha_{s^{-1}}(x)(\xi(s) - \xi(s_0)) \| + \| (\alpha_{s^{-1}}(x) - \alpha_{s_0^{-1}}(x))\xi(s_0) \| \\ &\leq \|x\| \, \| \xi(s) - \xi(s_0) \| + \| (\alpha_{s^{-1}}(x) - \alpha_{s_0^{-1}}(x))\xi(s_0) \| \end{aligned}$$

where the first term tends to zero as ξ is continuous and the second term because the action is continuous, that is $s \to \alpha_s(s)\xi_0$ is continuous, which is used here with $\xi_0 = \xi(s_0)$.

Finally

$$\begin{aligned}\|\xi_1\|^2 &= \int \|\xi_1(s)\|^2 ds \\ &= \int \|\alpha_{s^{-1}}(x)\xi(s)\|^2 ds \\ &\leq \int \|\alpha_{s^{-1}}(x)\|^2 \|\xi(s)\|^2 ds = \|x\|^2 \int \|\xi(s)\|^2 ds \\ &= \|x\|^2 \|\xi\|^2.\end{aligned}$$

Because of lemma 2.3 the following definition makes sense.

2.4 **Definition.** For every $x \in M$ we define a bounded operator $\pi(x)$ on $L_2(G, \mathcal{H})$ by

$$(\pi(x)\xi)(s) = \alpha_{s^{-1}}(x)\xi(s) \quad \text{for} \quad \xi \in C_c(G, \mathcal{H}).$$

If there are different actions around we will occasionally use π_α instead of π.

The crossed product will, among others, contain all the operators $\pi(x)$ with $x \in M$, therefore let us study them a little more.

2.5 **Proposition.** π is a faithful normal *-representation of M in $L_2(G, \mathcal{H})$.

Proof. Using the fact that α_s is a *-automorphism of M for each $s \in G$, a straightforward calculation shows that π is a *-representation.

Let us show that π is faithful, therefore assume $x \in M$ and $\pi(x) = 0$. Take $\xi_0 \in \mathcal{H}$ and $f \in C_c(G)$ and let $\xi = \xi_0 \otimes f$. Then

$$\begin{aligned}0 = \langle \pi(x)\xi, \xi\rangle &= \int\langle(\pi(x)\xi)(s), \xi(s)\rangle ds = \int\langle\alpha_{s^{-1}}(x)\xi(s), \xi(s)\rangle ds \\ &= \int\langle\alpha_{s^{-1}}(x)f(s)\xi_0, f(s)\xi_0\rangle ds = \int\langle\alpha_{s^{-1}}(x)\xi_0, \xi_0\rangle |f(s)|^2 ds.\end{aligned}$$

Because this holds for all $f \in C_c(G)$, and because $\langle\alpha_{s^{-1}}(x)\xi_0, \xi_0\rangle$ is continuous in s, we have that $\langle \alpha_{s^{-1}}(x)\xi_0, \xi_0\rangle = 0$ for all s, in particular $\langle x\xi_0, \xi_0\rangle = 0$. Again this holds for all $\xi_0 \in \mathcal{H}$ so that $x = 0$. This proves that π is faithful.

To prove that π is normal, let $\{x_i\}_{i\in I}$ be a bounded, increasing net in M^+ with x as supremum. As π is a *-representation also $\pi(x_i)$ will be bounded and increasing and therefore will increase to some operator on $L_2(G, \mathcal{H})$, call it $\tilde{y}$. As $x_i \leq x$ for all i we also have $\pi(x_i) \leq \pi(x)$ so that $\tilde{y} \leq \pi(x)$. We must show $\tilde{y} = \pi(x)$.

First let $f \in C_c(G)$ and $\xi_0 \in \mathcal{H}$, then with $\xi = \xi_0 \otimes f$ we have as before

$$\langle \pi(x_i)\xi, \xi\rangle = \int\langle \alpha_{s^{-1}}(x_i)\xi_0, \xi_0\rangle |f(s)|^2 ds.$$

Now $s \to \langle \alpha_{s^{-1}}(x_i)\xi_0, \xi_0\rangle$ is a net of continuous positive functions, increasing to the function $s \to \langle \alpha_{s^{-1}}(x)\xi_0, \xi_0\rangle$ which is also continuous. By Dini's theorem [10] we have uniform convergence on compacta. In particular we have uniform convergence on the support of f, so also the integrals will converge. Hence $\langle \pi(x_i)\xi, \xi\rangle \to \langle \pi(x)\xi, \xi\rangle$. So we have $\langle \pi(x)\xi, \xi\rangle = \langle \tilde{y}\xi, \xi\rangle$. Now as $\tilde{y} \leq \pi(x)$ it follows from $\langle (\pi(x) - \tilde{y})\xi, \xi\rangle = 0$ that also $(\pi(x) - \tilde{y})\xi = 0$ or $\pi(x)\xi = \tilde{y}\xi$. This implies $\pi(x) = \tilde{y}$ because functions of the form $\xi = \xi_0 \otimes f$ with $\xi_0 \in \mathcal{H}$ and $f \in C_c(G)$ span a dense subspace.

It is mainly because of the definition of $\pi(x)$ that we must work with $L_2(G, \mathcal{H})$ instead of $\mathcal{H} \otimes L_2(G)$. It is good to keep in mind that roughly speaking $\pi(x)$ is the multiplication operator in $L_2(G, \mathcal{H})$ by the function $s \to \alpha_{s^{-1}}(x)$. Multiplication operators will play an important role in the next section. Before we continue let us consider an example.

2.6 **Example.** Assume here that G is a finite group with n elements $\{s_1, s_2, \ldots, s_n\}$. Then using the appropriate normalization of the Haar measure, $L_2(G, \mathcal{H})$ can be identified with the direct sum $\mathcal{H} \oplus \mathcal{H} \oplus \ldots \mathcal{H}$ of n copies of $\mathcal{H}$ by means of the isomorphism

$$\xi \in L_2(G, \mathcal{H}) \to (\xi(s_1), \xi(s_2) \ldots \xi(s_n)).$$

Then in matrix notation we get the following expression for $\pi(x)$:

$$\pi(x) = \begin{pmatrix} \alpha_{s_1^{-1}}(x) & 0 & \cdots & 0 \\ 0 & \alpha_{s_2^{-1}}(x) & \cdots & 0 \\ \vdots & \vdots & \vdots & \vdots \\ 0 & 0 & \cdots & \alpha_{s_n^{-1}}(x) \end{pmatrix} .$$

In the next section we will obtain more information about the relation of $\pi(x)$ with the tensor product structure of $L_2(G, \mathcal{H})$. Now we define a representation of G in $L_2(G, \mathcal{H})$. This will provide us with the second type of operators that will generate the crossed product.

2.7 **Definition.** For every $t \in G$ we define a bounded operator $\lambda(t)$ on $L_2(G, \mathcal{H})$ by

$$(\lambda(t)\xi)(s) = \xi(t^{-1}s)$$

with $\xi \in C_c(G, \mathcal{H})$ and $s \in G$.

It follows easily from the invariance of the Haar measure that such an operator exists and is isometric.

2.8 **Proposition.** λ _is a strongly continuous unitary representation of_ G _in_ $L_2(G, \mathcal{H})$.

Proof. It is straightforward to verify that λ is indeed a representation of G and that $\lambda(t)$ is unitary for all $t \in G$. The proof of the strong continuity can either be obtained in a similar way as in the case $L_2(G)$, or can be obtained from the corresponding result in $L_2(G)$. Indeed, let $\xi = \xi_0 \otimes f$ with $\xi_0 \in \mathcal{H}$ and $f \in C_c(G)$. Then if we denote left translation by t^{-1} in $L_2(G)$ by λ_t we get

$$(\lambda(t)\xi)(s) = \xi(t^{-1}s) = f(t^{-1}s)\xi_0 = (\lambda_t f)(s)\xi_0 = (\xi_0 \otimes \lambda_t f)(s).$$

It follows that $\lambda(t) = 1 \otimes \lambda_t$ and the strong continuity in $L_2(G, \mathcal{H})$ follows from the one in $L_2(G)$ [11, p. 118].

The following relation shows that in the representation π the automorphisms α_t are unitarily implemented by the operators $\lambda(t)$.

2.9 **Lemma.** For all $x \in M$ and $t \in G$ we have $\lambda(t)\pi(x)\lambda(t)^* = \pi(\alpha_t(x))$.

Proof. Let $\xi \in C_c(G, \mathcal{H})$ and $s \in G$, then

$$(\lambda(t)\pi(x)\lambda(t)^*\xi)(s) = (\pi(x)\lambda(t)^*\xi)(t^{-1}s) = \alpha_{s^{-1}t}(x)(\lambda(t)^*\xi)(t^{-1}s)$$
$$= \alpha_{s^{-1}}(\alpha_t(x))\xi(s) = (\pi(\alpha_t(x))\xi)(s)$$

and the result follows.

It follows from this lemma that linear combinations of operators of the form $\pi(x)\lambda(t)$ with $x \in M$ and $t \in G$ form a *-algebra. Indeed, let $x, y \in M$ and $t, s \in G$ then

$$\begin{aligned}\pi(x)\lambda(t)\pi(y)\lambda(s) &= \pi(x)\lambda(t)\pi(y)\lambda(t)^*\lambda(t)\lambda(s)\\ &= \pi(x)\pi(\alpha_t(y))\lambda(ts)\\ &= \pi(x\alpha_t(y))\lambda(ts)\end{aligned}$$

and

$$(\pi(x)\lambda(t))^* = \lambda(t)^*\pi(x^*) = \lambda(t^{-1})\pi(x^*)\lambda(t^{-1})^*\lambda(t^{-1}) = \pi(\alpha_{t^{-1}}(x^*))\lambda(t^{-1}).$$

We now come to the definition of a crossed product.

2.10 **Definition.** The crossed product of M by the action α of G is the von Neumann algebra generated by the operators $\{\pi(x), \lambda(s) \mid x \in M, s \in G\}$ and is denoted by $M \otimes_\alpha G$. Because of the preceding it is the closure of the *-algebra of linear combinations of products $\pi(x)\lambda(s)$ with $x \in M$ and $s \in G$.

2.11 **Example.** Assume that G only has two elements $\{e, s\}$ where e is the identity. Then as in 2.6 the space $L_2(G, \mathcal{H})$ is identified with $\mathcal{H} \oplus \mathcal{H}$ and for $\pi(x)$ we get

$$\pi(x) = \begin{pmatrix} x & 0 \\ 0 & \alpha_s(x) \end{pmatrix}$$

as here $\alpha_e(x) = x$ and $\alpha_{s^{-1}}(x) = \alpha_s(x)$. Also

$$\lambda(s) = \begin{pmatrix} 0 & 1 \\ 1 & 0 \end{pmatrix}$$

and as $M \otimes_\alpha G = \{\pi(x) + \pi(y)\lambda(s) \mid x, y \in M\}$ we have in this case that the elements of $M \otimes_\alpha G$ are precisely those of the form

$$\begin{pmatrix} x & y \\ \alpha_s(y) & \alpha_s(x) \end{pmatrix}$$

with $x, y \in M$.

It is straightforward to check that those operators indeed form a *-algebra.

There are various other ways to construct the crossed product $M \otimes_\alpha G$. A particularly easy one to work with can be obtained in the case where α is implemented.

2.12 Proposition. *Assume that there is a strongly continuous unitary representation* $a : s \to a_s$ *of* G *in* $\mathcal{H}$ *such that* $\alpha_s(x) = a_s x a_s^*$ *for all* $x \in M$ *and* $s \in G$. *Then with the unitary operator* W *defined on* $L_2(G, \mathcal{H})$ *by* $(W\xi)(s) = a_s \xi(s)$ *for* $\xi \in C_c(G, \mathcal{H})$ *and* $s \in G$, *we obtain*

$$\pi(x) = W^*(x \otimes 1)W$$

and

$$\lambda(s) = W^*(a_s \otimes \lambda_s)W.$$

In particular $M \otimes_\alpha G$ *is spatially isomorphic to the von Neumann algebra in* $\mathcal{H} \otimes L_2(G)$ *generated by the operators* $\{x \otimes 1, a_s \otimes \lambda_s \mid x \in M, s \in G\}$ *(recall that* λ_s *is left translation by* s^{-1} *in* $L_2(G)$*).*

Proof. It is easily seen that W is well defined by $(W\xi)(s) = a_s\xi(s)$ for $\xi \in C_c(G, \mathcal{H})$, that it is unitary and that $(W^*\xi)(s) = a_s^*\xi(s)$. Now if $x \in M$ we can define an operator $\tilde{x}$ on $L_2(G, \mathcal{H})$ by $(\tilde{x}\xi)(s) = x\xi(s)$ for $\xi \in C_c(G, \mathcal{H})$. If ξ is of the form

$\xi_0 \otimes f$ with $\xi_0 \in \mathcal{H}$ and $f \in C_c(G)$ we get $x\xi(s) = xf(s)\xi_0 = f(s)x\xi_0 = (x\xi_0 \otimes f)(s)$.

Therefore $\tilde{x} = x \otimes 1$ and we obtain that $x \otimes 1$ maps $C_c(G, \mathcal{H})$ into $C_c(G, \mathcal{H})$ and that

$$((x \otimes 1)\xi)(s) = x\xi(s).$$

Then

$$(W^*(x \otimes 1)W\xi)(s) = a_s^*((x \otimes 1)W\xi)(s) = a_s^* x(W\xi)(s) = a_s^* x a_s \xi(s)$$
$$= \alpha_{s^{-1}}(x)\xi(s) = (\pi(x)\xi)(s)$$

proving $\pi(x) = W^*(x \otimes 1)W$.

Similarly $(a_s \otimes \lambda_s \xi)(t) = a_s \xi(s^{-1}t)$ for $\xi \in C_c(G, \mathcal{H})$ so that

$$(W^*(a_s \otimes \lambda_s)W\xi)(t) = a_t^*((a_s \otimes \lambda_s)W\xi)(t)$$
$$= a_t^* a_s (W\xi)(s^{-1}t)$$
$$= a_{t^{-1}s}\, a_{s^{-1}t}\, \xi(s^{-1}t)$$
$$= \xi(s^{-1}t)$$
$$= (\lambda(s)\xi)(t)$$

and

$$\lambda(s) = W^*(a_s \otimes \lambda_s)W.$$

In many situations it is much easier to work with the form $\{x \otimes 1,\ a_s \otimes \lambda_s \mid x \in M,\ s \in G\}$, as we will see e.g. when we treat the duality theorem in section 4. From propositions 2.5 and 2.8 and lemma 2.9 it follows that there is always a faithful normal representation of M in which the action is unitarily implemented in the above sense. This, together with the following proposition shows that in fact it is no restriction to assume that the action is unitarily implemented, and so that one can as well work with the simpler form above.

2.13 **Proposition.** *Let M and N be von Neumann algebras, and τ an isomorphism of M onto N. Suppose α and β are continuous actions of G on M and N respectively, related by $\tau(\alpha_t(x)) = \beta_t(\tau(x))$*

for all $x \in M$ and $t \in G$. Then there is an isomorphism $\tilde{\tau}$ of $M \otimes_\alpha G$ onto $N \otimes_\beta G$.

The proof of this proposition can be obtained using the structure of isomorphisms (see [16]). We do not give this proof here as it will follow easily from later results in section 3.

3. THE COMMUTATION THEOREM FOR CROSSED PRODUCTS

At the end of the previous section we obtained another expression for the crossed product, more related to the tensor product structure of $L_2(G, \mathcal{H}) = \mathcal{H} \otimes L_2(G)$. In this section we will get still another characterization; namely we will show that $M \otimes_\alpha G$ is the fixed point algebra in $M \otimes \mathcal{B}(L_2(G))$ for a certain action of G on $M \otimes \mathcal{B}(L_2(G))$. In the case where α is spatial this yields a characterization of the commutant of $M \otimes_\alpha G$. We first prove the following.

3.1 Lemma. $M \otimes_\alpha G \subseteq M \otimes \mathcal{B}(L_2(G))$.

Proof. We have seen that $\lambda(t) = 1 \otimes \lambda_t$ so that obviously $\lambda(t) \in M \otimes \mathcal{B}(L_2(G))$. So the proof will be complete if we show that $\pi(x)$ and $x' \otimes 1$ commute for every $x \in M$ and $x' \in M'$ as this will imply that $\pi(x) \in M \otimes \mathcal{B}(L_2(G))$. Now as in the proof of proposition 2.12 we have that $(x' \otimes 1)\xi \in C_c(G, \mathcal{H})$ for each $\xi \in C_c(G, \mathcal{H})$ and that $((x' \otimes 1)\xi)(s) = x'\xi(s)$. Thus

$$\begin{aligned}(\pi(x)(x' \otimes 1)\xi)(s) &= \alpha_{s^{-1}}(x)x'\xi(s)\\ &= x'\alpha_{s^{-1}}(x)\xi(s)\\ &= ((x' \otimes 1)\pi(x)\xi)(s)\end{aligned}$$

so that $\pi(x)(x' \otimes 1) = (x' \otimes 1)\pi(x)$.

Now we define an action θ of G on $M \otimes \mathcal{B}(L_2(G))$.

3.2 Notations. As before let λ_t denote left translation on $L_2(G)$, so $(\lambda_t f)(s) = f(t^{-1}s)$ for $f \in L_2(G)$. We will also use right translation, so define ρ_t for each $t \in G$ by $(\rho_t f)(s) = \Delta(t)^{\frac{1}{2}} f(st)$ where

$f \in L_2(G)$ and Δ is the modular function of G. Then also ρ is a continuous unitary representation of G in $L_2(G)$ [see 11, p. 118]. Denote by $\mathrm{ad}\rho_t$ the mapping $a \to \rho_t a \rho_t^*$ from $\mathcal{B}(L_2(G))$ into itself. As we remarked before $t \to \mathrm{ad}\rho_t$ will be a continuous action of G on $\mathcal{B}(L_2(G))$. Then define θ_t on $M \otimes \mathcal{B}(L_2(G))$ by $\theta_t = \alpha_t \otimes \mathrm{ad}\rho_t$. It is fairly easy to see that θ is again an action of G on $M \otimes \mathcal{B}(L_2(G))$. The continuity of θ follows from the following proposition.

3.3 Proposition. Let α and β be continuous actions of G on von Neumann algebras M and N respectively, then $\gamma_t = \alpha_t \otimes \beta_t$ defines an action γ of G on $M \otimes N$ which is also continuous.

Proof. First remark that γ_t is defined by $\gamma_t(m \otimes n) = \alpha_t(m) \otimes \beta_t(n)$ and that γ_t is indeed a *-automorphism of $M \otimes N$ [14, p. 67]. It is also immediate that γ is an action of G. Because M and N are considered with their strong topologies instead of norm topologies the continuity of γ is not obvious but requires an argument. Suppose first that α and β are spatial. So, if M and N act in $\mathcal{H}$ and $\mathcal{K}$ respectively, there are continuous unitary representations a and b of G on $\mathcal{H}$ and $\mathcal{K}$ respectively such that $\alpha_t(x) = a_t x a_t^*$ and $\beta_t(y) = b_t y b_t^*$ for $x \in M$ and $y \in N$. Then $u_t = a_t \otimes b_t$ is well defined and u is again a continuous unitary representation of G on $\mathcal{H} \otimes \mathcal{K}$. Indeed $t \to u_t \xi$ will be continuous on linear combinations of vectors of the form $\xi_1 \otimes \xi_2$ with $\xi_1 \in \mathcal{H}$ and $\xi_2 \in \mathcal{K}$ and they are dense. Now $\alpha_t(x) \otimes \beta_t(y) = u_t(x \otimes y)u_t^*$ so that $\gamma_t(z) = u_t z u_t^*$ for all $z \in M \otimes N$ and the continuity of γ follows.

In the general case there are normal faithful *-representations π_1 of M in $\mathcal{H}_1$ and π_2 of N in $\mathcal{H}_2$, and continuous unitary representations a and b of G in $\mathcal{H}_1$ and $\mathcal{H}_2$ respectively such that $\pi_1(\alpha_t(x)) = a_t \pi_1(x) a_t^*$ and $\pi_2(\beta_t(x)) = b_t \pi_2(x) b_t^*$. Again if $u_t = a_t \otimes b_t$ and $\pi = \pi_1 \otimes \pi_2$ then $\pi(\gamma_t(z)) = u_t \pi(z) u_t^*$. So $t \to \pi(\gamma_t(z))$ and hence $t \to \gamma_t(z)$ is strongly continuous, because also π is a normal faithful *-representation of $M \otimes N$.

Let us now come back to the action θ of G on $M \otimes \mathcal{B}(L_2(G))$. It is obvious that $\theta_t(\lambda(s)) = \lambda(s)$ for all $t, s \in G$ as left and right translations commute. As we will see also $\theta_t(\pi(x)) = \pi(x)$ for all $t \in G$ and $x \in M$ so that $\theta_t(\tilde{x}) = \tilde{x}$ for all $\tilde{x} \in M \otimes_\alpha G$. The main difficulty however is to show that any fixed point in $M \otimes \mathcal{B}(L_2(G))$ is in $M \otimes_\alpha G$.

In our approach we will essentially work with the map $\tilde{x} \to \int \theta_t(\tilde{x})dt$. This is no problem when the group is compact. In fact, provided the Haar measure is normalised, this gives a normal projection onto the fixed points. Because G need not be compact we will have to use the appropriate approximation. We start with a general basic lemma.

3.4 **Lemma.** Let $\tilde{x} \in M \otimes \mathcal{B}(L_2(G))$ and K a compact set in G, then $\int_K \theta_t(\tilde{x})dt$ is well-defined in the σ-weak topology on $M \otimes \mathcal{B}(L_2(G))$ and the map

$$\tilde{x} \to \int_K \theta_t(\tilde{x})dt$$

is σ-weakly continuous.

Proof. Denote for a moment $\tilde{M} = M \otimes \mathcal{B}(L_2(G))$ and $\tilde{\mathcal{H}} = \mathcal{H} \otimes L_2(G)$. Let $\xi, \eta \in \tilde{\mathcal{H}}$, then $t \to \langle \theta_t(\tilde{x})\xi, \eta\rangle$ is continuous and we can define $\int_K \langle \theta_t(\tilde{x})\xi, \eta\rangle dt$. This expression is clearly linear in ξ and conjugate linear in η and because

$$\left| \int_K \langle \theta_t(\tilde{x})\xi, \eta\rangle dt \right| \leq \int_K |\langle \theta_t(\tilde{x})\xi, \eta\rangle| dt \leq \|\tilde{x}\| \, \|\xi\| \, \|\eta\| \int_K dt$$

we get the existence of a bounded operator $\tilde{y}$ such that

$$\langle \tilde{y}\xi, \eta\rangle = \int_K \langle \theta_t(\tilde{x})\xi, \eta\rangle dt \quad \text{for all } \xi, \eta \in \tilde{\mathcal{H}}.$$

Now if $\tilde{z} \in \tilde{M}'$, the commutant of $\tilde{M}$, then

$$\langle \theta_t(\tilde{x})\tilde{z}\xi, \eta\rangle = \langle \tilde{z}\theta_t(\tilde{x})\xi, \eta\rangle = \langle \theta_t(\tilde{x})\xi, \tilde{z}^*\eta\rangle$$

so that also

$$\langle \tilde{y}\,\tilde{z}\xi, \eta\rangle = \langle \tilde{y}\xi, \tilde{z}^*\eta\rangle = \langle \tilde{z}\tilde{y}\xi, \eta\rangle .$$

Hence $\tilde{y}\tilde{z} = \tilde{z}\tilde{y}$ and $\tilde{y} \in \tilde{M}$.

Next let ϕ be any σ-weakly continuous linear functional on $\tilde{M}$, then ϕ is approximated in norm by a sequence ϕ_n of linear combinations of functionals of the type $\tilde{x} \to \langle \tilde{x}\xi, \eta \rangle$. For any ϕ_n we have

$$\phi_n(\tilde{y}) = \int_K \phi_n(\theta_t(\tilde{x}))dt$$

so that

$$\begin{aligned} |\phi(\tilde{y}) - \int_K \phi(\theta_t(x)dt| &\leq |\phi(\tilde{y}) - \phi_n(\tilde{y})| + \int_K |\phi(\theta_t(\tilde{x})) - \phi_n(\theta_t(\tilde{x}))|dt \\ &\leq \|\phi - \phi_n\| \|\tilde{y}\| + \|\phi - \phi_n\| \|\tilde{x}\| \int_K dt \end{aligned}$$

so that also $\phi(\tilde{y}) = \int_K \phi(\theta_t(\tilde{x}))dt$ and the integral exists in the σ-weak topology.

It remains to show that the map $x \to \int_K \theta_t(\tilde{x})dt$ is normal because that implies σ-weak continuity. So let $\{\tilde{x}_i\}$ be an increasing net of positive elements in $\tilde{M}$ with supremum $\tilde{x}$. For any $\xi \in \tilde{\mathcal{H}}$ we will have that $\langle \theta_t(\tilde{x}_i)\xi, \xi \rangle \nearrow \langle \theta_t(\tilde{x})\xi, \xi \rangle$ for each t and as in the proof of proposition 2.5 we can use Dini's theorem to get that

$$\langle (\int_K \theta_t(\tilde{x}_i)dt)\xi, \xi \rangle = \int_K \langle \theta_t(\tilde{x}_i)\xi, \xi \rangle dt \to \int_K \langle \theta_t(\tilde{x})\xi, \xi \rangle dt = \langle (\int_K \theta_t(\tilde{x})dt)\xi, \xi \rangle$$

Therefore $\int_K \theta_t(\tilde{x}_i)dt \nearrow \int_K \theta_t(\tilde{x})dt$ and the proof is complete.

To get into the fixed points for θ, it really ought to be possible to integrate over the whole group G. In general the integral over G of $\theta_t(\tilde{x})$ will not be defined. However because of the very special form of θ here, it will be defined for enough elements in $M \otimes \mathcal{B}(L_2(G))$ to allow an approximation procedure. The main reason for this is that ρ_t is a translation, so that also $\mathrm{ad}\rho_t$ acts as a translation on multiplication operators, and similarly θ_t. Therefore multiplication operators provide us with the right objects in our approximation procedure.

3.5 **Notation.** Let $f \in C_c(G)$, then we denote by m_f the multiplication in $L_2(G)$ by the function f, so $(m_f g)(s) = f(s)g(s)$ for all $g \in C_c(G)$. Similarly we denote by $m(f)$ multiplication by f in $L_2(G, \mathcal{H})$, so $(m(f)\xi)(s) = f(s)\xi(s)$ for all $\xi \in C_c(G, \mathcal{H})$. Also here we will have that $m(f) = 1 \otimes m_f$. Indeed let $\xi_0 \in \mathcal{H}$ and $g \in C_c(G)$ then

$$(m(f)(\xi_0 \otimes g))(s) = f(s)g(s)\xi_0 = (m_f g)(s)\xi_0 = (\xi_0 \otimes m_f g)(s).$$

The following lemma is crucial.

3.6 **Lemma.** _If_ $f \in C_c(G)$ _then_ $\int \theta_t(m(f))dt = \int f(t)dt \cdot 1$ _in the_ _σ-weak topology where_ 1 _is the identity operator on_ $L_2(G, \mathcal{H})$.

Proof. Let us first see what $\theta_t(m(f))$ does. By definition $\theta_t(m(f)) = 1 \otimes \rho_t m_f \rho_t^*$. Take $g \in C_c(G)$, then

$$\begin{aligned}(\rho_t m_f \rho_t^* g)(s) &= \Delta(t)^{\frac{1}{2}}(m_f \rho_t^* g)(st) \\ &= \Delta(t)^{\frac{1}{2}} f(st)(\rho_t^* g)(st) \\ &= f(st)g(s).\end{aligned}$$

This implies that $(\theta_t(m(f))\xi)(s) = f(st)\xi(s)$ for any $\xi \in C_c(G, \mathcal{H})$. Now let $h \in C_c(G)$, then $\theta_t(m(f))m(h)$ will be multiplication by the function $\phi_t(s) = f(st)h(s)$. Because h and f have compact support, and a product of compact sets is compact, the function ϕ_t will be identically zero for t outside some compact set, and hence $t \to \theta_t(m(f))m(h)$ will have compact support. Then in a similar way as in lemma 3.4, $\int \theta_t(m(f))m(h)dt$ exists in the σ-weak topology. Then let $\xi, \eta \in C_c(G, \mathcal{H})$, we get

$$\begin{aligned}\int \langle \theta_t(m(f))m(h)\xi, \eta \rangle dt &= \int(\int \langle f(st)h(s)\xi(s), \eta(s)\rangle ds)dt \\ &= \int(\langle h(s)\xi(s), \eta(s)\rangle \int f(st)dt)ds \\ &= \int f(t)dt \cdot \int \langle h(s)\xi(s), \eta(s)\rangle ds \\ &= \int f(t)dt \langle m(h)\xi, \eta\rangle.\end{aligned}$$

(We may use Fubini here as we are only integrating over compact sets.) Hence

$$\int \theta_t(m(f))m(h)dt = \int f(t)dt \cdot m(h). \qquad (*)$$

Next let $\xi \in L_2(G)$, then there is a sequence of compact sets K_n, such that K_n increases to $\cup K_n$ containing the support of ξ. Then choose an increasing sequence $h_n \in C_c(G)$ with $0 \le h_n \le 1$ and $h_n = 1$ on K_n. If now $f \ge 0$ we will have $\langle \theta_t(m(f))m(h_n)\xi, \xi\rangle \ge 0$ as $\theta_t(m(f))$ and $m(h_n)$ are commuting positive operators. Because h_n is increasing, also $\langle\theta_t(m(f))m(h_n)\xi, \ \xi\rangle$ will be increasing. Finally

$$\|m(h_n)\xi - \xi\|^2 = \int \|h_n(s)\xi(s) - \xi(s)\|^2 ds \le \int_{G\setminus K_n} \|\xi(s)\|^2 ds \to 0$$

as K_n increases to $\cup K_n$ containing the support of ξ. Therefore

$$\langle\theta_t(m(f))m(h_n)\xi, \ \xi\rangle \nearrow \langle\theta_t(m(f))\xi, \ \xi\rangle$$

while also $\langle m(h_n)\xi, \ \xi\rangle \nearrow \langle\xi, \ \xi\rangle$. From (*) we had already that

$$\int\langle\theta_t(m(f))m(h_n)\xi, \ \xi\rangle dt = \int f(t)dt\langle m(h_n)\xi, \ \xi\rangle$$

and then by the monotone convergence theorem we have

$$\int\langle\theta_t(m(f))\xi, \ \xi\rangle dt = \int f(t)dt\langle\xi, \ \xi\rangle .$$

Now as any positive normal functional ϕ can be obtained as a limit of an increasing sequence of positive linear combinations of vector states, again by the monotone convergence theorem it follows that

$$\int \phi(\theta_t(m(f)))dt = \int f(t)dt \, \phi(1).$$

It then follows by linearity that the integral $\int \theta_t(m(f))dt$ exists in the σ-weak topology and that it equals $\int f(t)dt \cdot 1$. This completes the proof.

3.7 Lemma. Let f and $g \in C_c(G)$, then for all $\tilde{x} \in M \otimes \mathcal{B}(L_2(G))$ we have that $\int \theta_t(m(f)\tilde{x}\, m(g))dt$ is well defined in the σ-weak topology, and the expression is σ-weakly continuous in $\tilde{x}$.

Proof. Let $\xi, \eta \in L_2(G, \mathcal{H})$, then

$$|\langle\theta_t(m(f)\tilde{x}\, m(g))\xi, \ \eta\rangle| \le \|\tilde{x}\| \, \|\theta_t(m(g))\xi\| \, \|\theta_t(m(\bar{f}))\eta\|$$

where as usual $\bar{f}(s) = \overline{f(s)}$. Now from lemma 3.6 we know that

$$\|\theta_t(m(g))\xi\|^2 = \langle\theta_t(m(\bar{g}g))\xi, \ \xi\rangle$$

is integrable over t so that $t \to \|\theta_t(m(g))\xi\|$ is an L_2-function. Similarly $t \to \|\theta_t(m(\bar{f}))\eta\|$, and because the product of two L_2-functions is L_1 we have that $t \to \langle\theta_t(m(f)\tilde{x}\, m(g))\xi, \ \eta\rangle$ is integrable. Furthermore

$$|\int\langle\theta_t(m(f)\tilde{x}\,m(g))\xi,\ \eta\rangle dt| \le \|\tilde{x}\|\int\|\theta_t(m(g))\xi\|\,\|\theta_t(m(\bar{f}))\eta\|\,dt$$
$$\le \|\tilde{x}\|(\int\|\theta_t(m(g))\xi\|^2 dt)^{\frac{1}{2}}(\int\|\theta_t(m(\bar{f}))\eta\|^2 dt)^{\frac{1}{2}}$$
$$= \|\tilde{x}\|(\int|g(t)|^2 dt)^{\frac{1}{2}}(\int|f(t)|^2 dt)^{\frac{1}{2}}\|\xi\|\,\|\eta\|$$
$$= \|\tilde{x}\|\,\|g\|\,\|f\|\,\|\xi\|\,\|\eta\|.$$

Then as in lemma 3.4 there is a $\tilde{y} \in M \otimes \mathcal{B}(L_2(G))$ such that

$$\langle\tilde{y}\xi,\ \eta\rangle = \int\langle\theta_t(m(f)\tilde{x}\,m(g))\xi,\ \eta\rangle dt$$

for all $\xi,\ \eta \in L_2(G,\ \mathcal{H})$.

Next suppose $g = \bar{f}$ and $\tilde{x} \ge 0$, then as in the proof of lemma 3.6 we can use the monotone convergence theorem to get that

$$\phi(\tilde{y}) = \int\phi(\theta_t(m(f)\tilde{x}\,m(\bar{f})))dt$$

for any positive normal functional ϕ. Then by polarization also $\phi(\tilde{y}) = \int\phi(\theta_t(m(f)\tilde{x}\,m(g)))dt$ in general.

Finally to prove σ-weak continuity, let $\xi \in C_c(G,\ \mathcal{H})$, then

$$t \to \langle\theta_t(m(f)\tilde{x}\,m(\bar{f}))\xi,\ \xi\rangle$$

has compact support as $(\theta_t(m(\bar{f}))\xi)(s) = \bar{f}(st)\xi(s)$ and f and ξ have compact support. It then follows from lemma 3.4 that

$$\tilde{x} \to \int\langle\theta_t(m(f)\tilde{x}\,m(\bar{f})\xi,\ \xi\rangle dt$$

is normal. As in the proof of proposition 2.5 this is sufficient to conclude that

$$\tilde{x} \to \int\theta_t(m(f)\tilde{x}\,m(\bar{f}))dt$$

is normal and therefore σ-weakly continuous. The result then follows by polarization.

3.8 Lemma. _If_ $x \in M$ _and_ $f \in C_c(G)$ _then_

$$\int\theta_t(x \otimes m_f)dt = \int f(t)\pi(\alpha_t(x))dt$$

in the σ-_weak topology._

Proof. The first integral exists because of lemma 3.7 while the second can be shown to exist as in lemma 3.4. Take $\xi \in C_c(G, \mathcal{H})$, then as before

$$t \to \langle \theta_t(x \otimes m_f)\xi, \xi\rangle$$

has compact support and

$$\begin{aligned}\int\langle \theta_t(x \otimes m_f)\xi, \xi\rangle dt &= \iint \langle \alpha_t(x) f(st)\xi(s), \xi(s)\rangle ds\, dt \\ &= \iint f(st)\langle \alpha_t(x)\xi(s), \xi(s)\rangle dt\, ds \\ &= \iint f(u)\langle \alpha_{s^{-1}}(\alpha_u(x))\xi(s), \xi(s)\rangle du\, ds \\ &= \iint f(u)\langle (\pi(\alpha_u(x))\xi)(s), \xi(s)\rangle ds\, du \\ &= \int f(u)\langle \pi(\alpha_u(x))\xi, \xi\rangle du.\end{aligned}$$

The calculation is quite similar as in lemma 3.6. Also here we can use Fubini's theorem as we are always integrating on compact sets.

This lemma shows already that $\theta_s(\pi(x)) = \pi(x)$ for all $x \in M$ and $s \in G$. Indeed, apply θ_s to the equality in lemma 3.8. This can be done as the integrals exist in the σ-weak topology and $*$-automorphisms are σ-weakly continuous.

So $\int \theta_s(\theta_t(x \otimes m_f))dt = \int f(t)\theta_s(\pi(\alpha_t(x)))dt$. But $\int \theta_s(\theta_t(x \otimes m_f))dt = \int \theta_{st}(x \otimes m_f)dt = \int \theta_t(x \otimes m_f)dt$ so that $\int f(t)\theta_s(\pi(\alpha_t(x)))dt = \int f(t)\pi(\alpha_t(x))dt$. This holds for all $f \in C_c(G)$ so that $\theta_s(\pi(\alpha_t(x))) = \pi(\alpha_t(x))$ for all s and for all t. Hence $\theta_s(\pi(x)) = \pi(x)$.

Proposition 2.13 can now also be proved using this result. We will do this at the end of this section. We should remark that lemma 3.8 can be obtained somewhat more directly but we need the different steps anyway in what follows.

So we have $\theta_s(\pi(x)) = \pi(x)$ and we had already that $\theta_s(\lambda(t)) = \lambda(t)$. Hence $\theta_s(\tilde{x}) = \tilde{x}$ for all $\tilde{x} \in M \otimes_\alpha G$. We now proceed to show that actually any fixed point is in $M \otimes_\alpha G$.

3.9 **Lemma.** <u>For all</u> $f, g \in C_c(G)$ <u>we have that</u>

$$\int \theta_t(m(f)\tilde{x}\, m(g))dt \in M \otimes_\alpha G \text{ for all } \tilde{x} \in M \otimes \mathcal{B}(L_2(G)).$$

Proof. First take $\tilde{x}$ of the form $x \otimes m_h \lambda_s$ with $x \in M$, $s \in G$ and $h \in C_c(G)$. Then

$$\begin{aligned}\theta_t(m(f)\tilde{x}\,m(g)) &= \theta_t(x \otimes m_f m_h \lambda_s m_g) \\ &= \theta_t(x \otimes m_\phi \lambda_s) \\ &= \theta_t(x \otimes m_\phi)\lambda(s)\end{aligned}$$

where $\phi(u) = f(u)h(u)g(s^{-1}u)$.

In any case, by lemma 3.8 we get

$$\int \theta_t(m(f)\tilde{x}\,m(g))dt = \int \phi(t)\pi(\alpha_t(x))\lambda(s)dt \in M \otimes_\alpha G.$$

Now the operators $m_h \lambda_s$ with $h \in C_c(G)$ and $s \in G$ span a σ-weakly dense subalgebra of $\mathcal{B}(L_2(G))$ (see appendix B) so that by the σ-weak continuity of the expression in $\tilde{x}$ (lemma 3.7) we obtain the present lemma.

If the group G is compact we would get the result here already. Since with $f = g = 1$ we would have that $\int \theta_t(\tilde{x})dt \in M \otimes_\alpha G$ for all $\tilde{x} \in M \otimes \mathcal{B}(L_2(G))$, and if $\tilde{x}$ is a fixed point we would get $\tilde{x} \in M \otimes_\alpha G$. The general proof that follows is entirely based on the same principle but uses an approximation argument. We need one more lemma.

3.10 **Lemma.** _Let_ $h \in C_c(G)$ _and_ K _compact. Then the function_ ϕ_K _defined by_ $\phi_K(s) = \int_K h(st)dt$ _is again in_ $C_c(G)$. _Moreover_ $m(\phi_K) = \int_K \theta_t(m(h))dt$ _and_ $m(\phi_K)$ _converges to_ $\int h(t)dt \cdot 1$ _in the_ σ-_weak topology when_ K _increases to_ G.

Proof. It is well known that ϕ_K is again continuous and has compact support [11]. Now take $\xi \in C_c(G, \mathcal{H})$, then

$$\begin{aligned}\int_K \langle \theta_t(m(h))\xi,\ \xi\rangle dt &= \int_K \int \langle h(st)\xi(s),\ \xi(s)\rangle ds\, dt \\ &= \int \langle (\int_K h(st)dt)\xi(s),\ \xi(s)\rangle ds \\ &= \int \langle \phi_K(s)\xi(s),\ \xi(s)\rangle ds \\ &= \langle m(\phi_K)\xi,\ \xi\rangle.\end{aligned}$$

Finally assume $h \geq 0$ and let $\xi \in C_c(G, \mathcal{H})$. Because $t \to \langle \theta_t(m(h))\xi, \xi \rangle$ has compact support, obviously

$$\langle (\int_K \theta_t(m(h))dt)\xi, \xi \rangle \nearrow \langle (\int \theta_t(m(h))dt)\xi, \xi \rangle = \int h(t)dt \langle \xi, \xi \rangle .$$

And because $\int_K \theta_t(m(h))dt \leq \int \theta_t(m(h))dt = \int h(t)dt \cdot 1$ we get that

$$\int_K \theta_t(m(h))dt \nearrow \int \theta_t(m(h))dt.$$

The general case follows by polarization.

We now prove the main theorem in this section.

3.11 Theorem. $M \otimes_\alpha G = \{\tilde{x} \in M \otimes \mathcal{B}(L_2(G)) \,|\, \theta_t(\tilde{x}) = \tilde{x}, \forall t \in G\}$.

Proof. We have shown already that $\theta_t(\tilde{x}) = \tilde{x}$ for all $t \in G$ and $\tilde{x} \in M \otimes_\alpha G$. Conversely take a $\tilde{x} \in M \otimes \mathcal{B}(L_2(G))$ such that $\theta_t(\tilde{x}) = \tilde{x}$. We will now show that $\tilde{x} \in M \otimes_\alpha G$. Fix a function $f \in C_c(G)$ and normalize it so that $\int f(t)dt = 1$ and define ϕ_K as in the previous lemma for any compact set K in G by $\phi_K(s) = \int_K f(st)dt$. By the same lemma $m(\phi_K) = \int_K \theta_t(m(f))dt$. Put

$$\tilde{x}_K = \int \theta_t(m(\phi_K)\tilde{x}\, m(f))dt.$$

Then from lemma 3.9 we know that $\tilde{x}_K \in M \otimes_\alpha G$. We will now show that for any $h \in C_c(G)$ we have $\tilde{x}_K m(h) \to \tilde{x}\, m(h)$ and $m(h)\tilde{x}_K \to m(h)\tilde{x}$. Then it will be easy to conclude that $\tilde{x} \in M \otimes_\alpha G$. So choose $h \in C_c(G)$, then

$$\tilde{x}_K m(h) = \int \theta_t(m(\phi_K)\tilde{x}\, m(f))m(h)dt = \int \theta_t(m(\phi_K))\tilde{x}\, \theta_t(m(f))m(h)dt.$$

Now as before $t \to \theta_t(m(f))m(h)$ has compact support, independent of K, so we are only integrating over a compact set, and then by lemma 3.4 we have σ-weak continuity. Then if K increases, $m(\phi_K) \to 1$ by lemma 3.10 as f is normalized such that $\int f(t)dt = 1$. Then

$$\tilde{x}_K m(h) \to \int \theta_t(1)\tilde{x}\, \theta_t(m(f))m(h)dt = \tilde{x} \int \theta_t(m(f))m(h)dt = \tilde{x}\, m(h).$$

To show that $m(h)\tilde{x}_K \to m(h)\tilde{x}$ is less immediate:

$$
\begin{aligned}
m(h)\tilde{x}_K &= \int m(h)\theta_t(m(\phi_K))\tilde{x}\,\theta_t(m(f))dt \\
&= \int(\int_K m(h)\theta_{ts}(m(f))\tilde{x}\,\theta_t(m(f))ds)dt \\
&= \int_K(\int m(h)\theta_{ts}(m(f))\tilde{x}\,\theta_t(m(f))dt)ds \\
&= \int_K \int \Delta(s)^{-1} m(h)\theta_u(m(f))\tilde{x}\,\theta_{us^{-1}}(m(f))du\,ds \\
&= \int \int_K \Delta(s^{-1})m(h)\theta_u(m(f))\tilde{x}\,\theta_{us^{-1}}(m(f))ds\,du \\
&= \int\int_{K^{-1}} m(h)\theta_u(m(f))\tilde{x}\,\theta_{us}(m(f))ds\,du \\
&= \int m(h)\theta_u(m(f))\tilde{x}\,\theta_u(m(\phi_{K^{-1}}))du.
\end{aligned}
$$

Again we can use Fubini as we are only integrating over compact sets. Similarly as for $\tilde{x}_K m(h)$ also here we are integrating u only over a compact set, namely the support of $u \to m(h)\theta_u(m(f))$ which is independent of K. Then as K increases, also K^{-1} will do so and $m(\phi_{K^{-1}}) \to 1$. Then as for $\tilde{x}_K m(h)$ also here

$$
m(h)\tilde{x}_K \to m(h)\tilde{x}.
$$

Now by similar methods as in [12] we can show that $\tilde{x} \in M \otimes_\alpha G$. Indeed let $\tilde{y} \in (M \otimes_\alpha G)'$, then

$$
\begin{aligned}
m(h)\tilde{y}\,\tilde{x}\,m(h) &= \lim_K m(h)\tilde{y}\,\tilde{x}_K\,m(h) \\
&= \lim_K m(h)\tilde{x}_K\,\tilde{y}\,m(h) \\
&= m(h)\tilde{x}\,\tilde{y}\,m(h)
\end{aligned}
$$

where we have used that $\tilde{x}_K \in M \otimes_\alpha G$. Finally let $m(h) \nearrow 1$ and we get $\tilde{y}\,\tilde{x} = \tilde{x}\,\tilde{y}$. This completes the proof.

To see more clearly what the above method has to do with techniques in [12], remark that $m(h)\tilde{x}_K \to m(h)\tilde{x}$ implies that $\tilde{x}_K^* m(h)^* \to \tilde{x}^* m(h)^*$. Then take $\tilde{x} = \tilde{x}^*$ and $h = \bar{h}$ so that not only $\tilde{x}_K m(h) \to \tilde{x}\,m(h)$ but also $\tilde{x}_K^* m(h) \to \tilde{x}\,m(h)$. Then if $\xi \in C_c(G, \mathcal{H})$ and $h = 1$ on the support of ξ so that $m(h)\xi = \xi$ we would get $\tilde{x}_K \xi \to \tilde{x}\xi$ and $\tilde{x}_K^* \xi \to \tilde{x}\xi$. In particular $\tilde{x}\xi \in \overline{(M \otimes_\alpha G)}_s\,\xi$. The fact is that this is now true for a dense set of vectors.

Now it is obvious to obtain a characterization of the commutant of $M \otimes_\alpha G$ if α is spatial.

3.12 Theorem. If α is spatial, that is if there exists a continuous unitary representation $a : s \to a_s$ of G in $\mathcal{H}$ such that $\alpha_s(x) = a_s x a_s^*$ then

$$(M \otimes_\alpha G)' = \{x' \otimes 1,\ a_s \otimes \rho_s \mid x' \in M',\ s \in G\}''.$$

Proof. By theorem 3.10 we have

$$M \otimes_\alpha G = \{\tilde{x} \in M \otimes \mathcal{B}(L_2(G)) \mid \theta_t(\tilde{x}) = \tilde{x},\ \forall t \in G\}.$$

Now $\theta_t(\tilde{x}) = \tilde{x}$ means $(a_t \otimes \rho_t)\tilde{x}(a_t \otimes \rho_t)^* = \tilde{x}$ as $\theta_t = \alpha_t \otimes \mathrm{ad}\,\rho_t$. Hence

$$M \otimes_\alpha G = M \otimes \mathcal{B}(L_2(G)) \cap \{a_s \otimes \rho_s \mid s \in G\}'$$

and $(M \otimes_\alpha G)' = ((M' \otimes 1) \cup \{a_s \otimes \rho_s \mid s \in G\})''$.

To finish this section, as promised we give a proof of proposition 2.13 of the end of the previous section. We give here a more precise formulation.

3.13 Proposition. Let M and N be von Neumann algebras acting in Hilbert spaces $\mathcal{H}$ and $\mathcal{K}$ respectively and let τ be an isomorphism of M onto N. Suppose α and β are continuous actions of G on M and N respectively, related by $\tau(\alpha_t(x)) = \beta_t(\tau(x))$ for all $x \in M$ and $t \in G$. Then $\tilde{\tau} = \tau \otimes 1$ is an isomorphism of $M \otimes_\beta G$ onto $N \otimes_\beta G$ such that $\tilde{\tau}(\pi_\alpha(x)) = \pi_\beta(\tau(x))$ where $x \in M$ and π_α and π_β are the representations as defined in 2.4 associated to α and β respectively.

Proof. As was proved in lemma 3.1 we have $M \otimes_\alpha G \subseteq M \otimes \mathcal{B}(L_2(G))$ and $N \otimes_\beta G \subseteq N \otimes \mathcal{B}(L_2(G))$ and $\tilde{\tau} : M \otimes \mathcal{B}(L_2(G)) \to N \otimes \mathcal{B}(L_2(G))$ so that $\tilde{\tau}$ is well defined on $M \otimes_\alpha G$.

From lemma 3.8 we know that

$$\int \alpha_t(x) \otimes \rho_t\, m_f\, \rho_t^*\, dt = \int f(t)\pi_\alpha(\alpha_t(x))dt$$

for any $x \in M$ and $f \in C_c(G)$.

Apply $\tilde{\tau} = \tau \otimes 1$ to this equation. Then

$$\int \tau(\alpha_t(x)) \otimes \rho_t\, m_f\, \rho_t^*\, dt = \int f(t)\tilde{\tau}(\pi_\alpha(\alpha_t(x)))dt$$
$$= \int \beta_t(\tau(x)) \otimes \rho_t\, m_f\, \rho_t^*\, dt$$
$$= \int f(t)\pi_\beta(\beta_t(\tau(x)))dt$$

by lemma 3.8 applied to N and β.

Hence $\int f(t)\pi_\beta(\beta_t(\tau(x)))dt = \int f(t)\tilde{\tau}(\pi_\alpha(\alpha_t(x)))dt$ for all $f \in C_c(G)$. So as before $\pi_\beta(\tau(x)) = \tilde{\tau}(\pi_\alpha(x))$. Obviously $\tilde{\tau}(\lambda(s)) = \lambda(s)$ and it follows that

$$\tilde{\tau}(M \otimes_\alpha G) = N \otimes_\beta G.$$

This completes the proof.

4. DUALITY

In this section we will assume that G is commutative, and we will consider the dual group $\hat{G}$ of G. If p is a character in $\hat{G}$, then $\langle s, p\rangle$ will denote the value of p in the point $s \in G$. $\hat{G}$ is again a locally compact group and we will write dp for the Haar measure on $\hat{G}$. We assume that the Haar measures on G and $\hat{G}$ are normalized in such a way that the Fourier transform becomes a unitary operator.

We will now define an action $\hat{\alpha}$ of $\hat{G}$ on $M \otimes_\alpha G$ in such a way that the crossed product $(M \otimes_\alpha G) \otimes_{\hat{\alpha}} \hat{G}$ can be shown to be isomorphic to $M \otimes \mathcal{B}(L_2(G))$.

4.1 Notation. For every character $p \in \hat{G}$ we denote by v_p the unitary operator on $L_2(G)$ defined by

$$(v_p f)(s) = \langle s, p\rangle f(s) \text{ for all } f \in C_c(G).$$

It is easily seen, that such an operator v_p exists and is unitary, and that v is a representation of $\hat{G}$ in $L_2(G)$. Because the topology on $\hat{G}$ is precisely the one of uniform convergence on compact sets we will have that $p \to v_p f$ is continuous for every $f \in C_c(G)$. This implies that v is a continuous representation.

We will now show that $(1 \otimes v_p)\tilde{x}(1 \otimes v_p)^* \in M \otimes_\alpha G$ for any $\tilde{x} \in M \otimes_\alpha G$.

4.2 **Lemma.** $v_p \lambda_s v_p^* = \overline{\langle s, p\rangle}\lambda_s$ for all $s \in G$ and $p \in \hat{G}$.

Proof. Let $f \in C_c(G)$, then

$$\begin{aligned}(v_p \lambda_s v_p^* f)(t) &= \overline{\langle t, p\rangle}(\lambda_s v_p^* f)(t) \\ &= \overline{\langle t, p\rangle}(v_p^* f)(s^{-1}t) \\ &= \overline{\langle t, p\rangle}\langle s^{-1}t, p\rangle f(s^{-1}t) \\ &= \overline{\langle s, p\rangle}(\lambda_s f)(t).\end{aligned}$$

4.3 **Lemma.** $(1 \otimes v_p)\pi(x)(1 \otimes v_p)^* = \pi(x)$ for all $x \in M$ and $p \in \hat{G}$.

Proof. If $\xi \in C_c(G, \mathcal{H})$, then as before $((1 \otimes v_p)\xi)(s) = \overline{\langle s, p\rangle}\,\xi(s)$ so that

$$\begin{aligned}((1 \otimes v_p)\pi(x)\xi)(s) &= \overline{\langle s, p\rangle}\alpha_{s^{-1}}(x)\xi(s) \\ &= \alpha_{s^{-1}}(x)\overline{\langle s, p\rangle}\,\xi(s) \\ &= (\pi(x)(1 \otimes v_p)\xi)(s).\end{aligned}$$

4.4 **Definition.** Define $\hat{\alpha}_p(\tilde{x}) = (1 \otimes v_p)\tilde{x}(1 \otimes v_p)^*$ for all $\tilde{x} \in M \otimes_\alpha G$ and $p \in \hat{G}$. As $M \otimes_\alpha G$ is generated by $\pi(x)$ and $\lambda(s)$ with $x \in M$ and $s \in G$, it follows from lemma 4.2 and 4.3 that $\tilde{\alpha}_p(\tilde{x}) \in M \otimes_\alpha G$. Hence $\hat{\alpha}$ is a continuous action of $\hat{G}$ on $M \otimes_\alpha G$. The action $\hat{\alpha}$ is called the dual action.

Of course $\hat{\alpha}$ is dependent on α because $M \otimes_\alpha G$ depends on α. The following proposition shows however that in some sense $\hat{\alpha}$ like $M \otimes_\alpha G$ is independent of the particular representation of α on M.

4.5 **Proposition.** Let M and N be von Neumann algebras and τ an isomorphism of M onto N. Suppose α and β are continuous actions of G on M and N respectively such that $\tau(\alpha_t(x)) = \beta_t(\tau(x))$ for all $t \in G$ and $x \in M$. Then $\tilde{\tau} = \tau \otimes 1$ is an isomorphism of $M \otimes_\alpha G$ onto $N \otimes_\beta G$ such that $\tilde{\tau}(\hat{\alpha}_p(\tilde{x})) = \hat{\beta}_p(\tilde{\tau}(\tilde{x}))$ for all $\tilde{x} \in M \otimes_\alpha G$ and $p \in \hat{G}$.

Proof. We showed already that $\tilde{\tau}$ is an isomorphism of $M \otimes_\alpha G$ onto $N \otimes_\beta G$. Now $\tilde{\tau}$ is also defined on $M \otimes \mathcal{B}(L_2(G))$ and it makes sense to write

$$\tilde{\tau}(\hat{\alpha}_p(\tilde{x})) = \tilde{\tau}(1 \otimes v_p)\tilde{\tau}(\tilde{x})\tilde{\tau}(1 \otimes v_p^*).$$

But $\tilde{\tau}(1 \otimes v_p) = (\tau \otimes 1)(1 \otimes v_p) = 1 \otimes v_p$ (where 1 denotes the identity in both M and N). So

$$\tilde{\tau}(\hat{\alpha}_p(\tilde{x})) = (1 \otimes v_p)\tilde{\tau}(\tilde{x})(1 \otimes v_p)^* = \hat{\beta}_p(\tilde{\tau}(\tilde{x})).$$

In what follows we will consider the crossed product $(M \otimes_\alpha G) \otimes_{\hat{\alpha}} \hat{G}$ of $M \otimes_\alpha G$ by the action $\hat{\alpha}$ of $\hat{G}$. It is a von Neumann algebra acting in $\mathcal{H} \otimes L_2(G) \otimes L_2(\hat{G})$. Because of proposition 4.5, also in studying this crossed product we may assume that α is spatial, since with the notations of proposition 4.5 we have that $\tau \otimes 1 \otimes 1$ will be an isomorphism of $(M \otimes_\alpha G) \otimes_{\hat{\alpha}} \hat{G}$ onto $(N \otimes_\beta G) \otimes_{\hat{\beta}} \hat{G}$.

We are now going to show that $(M \otimes_\alpha G) \otimes_{\hat{\alpha}} \hat{G}$ is isomorphic to $M \otimes \mathcal{B}(L_2(G))$. We will do this in a number of steps. We will assume that α is spatial which is no restriction for this purpose, and show that in that situation, $(M \otimes_\alpha G) \otimes_{\hat{\alpha}} \hat{G}$ is actually spatially isomorphic with $M \otimes \mathcal{B}(L_2(G)) \otimes 1$.

4.6 **Lemma.** <u>Denote</u> $\mathcal{R}_0 = (M \otimes_\alpha G) \otimes_{\hat{\alpha}} \hat{G}$, <u>and assume that</u> $a : s \to a_s$ <u>is a continuous unitary representation of</u> G <u>in</u> $\mathcal{H}$ <u>such that</u> $\alpha_t(x) = a_t x a_t^*$ <u>for all</u> $t \in G$ <u>and</u> $x \in M$. <u>Then</u> $\mathcal{R}_0$ <u>is spatially isomorphic to the von Neumann algebra</u> $\mathcal{R}_1$ <u>in</u> $\mathcal{H} \otimes L_2(G) \otimes L_2(\hat{G})$ <u>generated by the operators</u>

$$\{x \otimes 1 \otimes 1,\ a_s \otimes \lambda_s \otimes 1,\ 1 \otimes v_p \otimes \lambda_p \mid x \in M,\ s \in G,\ p \in \hat{G}\}$$

<u>where</u> λ_p <u>is left translation on</u> $L_2(\hat{G})$ <u>by</u> p^{-1}.

Proof. Because $\hat{\alpha}$ is spatial it will follow from proposition 2.12 that $(M \otimes_\alpha G) \otimes_{\hat{\alpha}} \hat{G}$ is spatially isomorphic to the von Neumann algebra generated by

$$(*) \qquad \{\tilde{x} \otimes 1,\ 1 \otimes v_p \otimes \lambda_p \mid \tilde{x} \in M \otimes_\alpha G,\ p \in \hat{G}\}.$$

Now also by this proposition $W(M \otimes_\alpha G)W^*$ is generated by $\{x \otimes 1,\ a_s \otimes \lambda_s \mid x \in M,\ s \in G\}$ where W is defined by $(W\xi)(s)=a_s\xi(s)$, $\xi \in C_c(G, \mathcal{H})$. Furthermore, W and $1 \otimes v_p$ commute (in a similar way as $\pi(x)$ and $1 \otimes v_p$ commute) so that if we apply $(W \otimes 1) \cdot (W^* \otimes 1)$ to the set of operators in $(*)$ we obtain the lemma.

4.7 **Lemma.** <u>$\mathcal{R}_1$ is spatially isomorphic to the von Neumann algebra $\mathcal{R}_2$ in $\mathcal{H} \otimes L_2(G) \otimes L_2(G)$ generated by</u>

$$\{x \otimes 1 \otimes 1,\ a_s \otimes \lambda_s \otimes 1,\ 1 \otimes v_p \otimes v_p \mid x \in M,\ s \in G,\ p \in \hat{G}\}.$$

Proof. We use the Fourier transform from $L_2(\hat{G})$ onto $L_2(G)$. So let $\mathcal{F}$ be the unitary operator from $L_2(\hat{G})$ onto $L_2(G)$ such that

$$(\mathcal{F}f)(t) = \int \overline{\langle t, p\rangle}\, f(p)dp \quad \text{for all } f \in C_c(\hat{G}).$$

Then to prove the lemma it will be sufficient to show that $\mathcal{F}\lambda_q\mathcal{F}^* = v_q$ for any $q \in \hat{G}$ because then we apply $(1 \otimes 1 \otimes \mathcal{F}) \cdot (1 \otimes 1 \otimes \mathcal{F}^*)$ to the operators generating $\mathcal{R}_1$ to get the operators generating $\mathcal{R}_2$. Therefore let $f \in C_c(\hat{G})$, then

$$(\mathcal{F}\lambda_q f)(t) = \int \overline{\langle t, p\rangle}(\lambda_q f)(p)dp = \int \overline{\langle t, p\rangle}\, f(q^{-1}p)dp = \int \overline{\langle t, qp\rangle} f(p)dp$$
$$= \overline{\langle t, q\rangle} \int \overline{\langle t, p\rangle} f(p)dp = \overline{\langle t, q\rangle}(\mathcal{F}f)(t) = (v_q\mathcal{F}f)(t).$$

For the last equality, to be precise, we should have defined v_q on $L_2(G)$ by $(v_q g)(s) = \overline{\langle s, q\rangle} g(s)$ for all $g \in L_2(G)$. Because also this gives a unitary operator, and $C_c(G)$ is dense, it follows that one can write $(v_q g)(s) = \overline{\langle s, q\rangle} g(s)$ also with our first definition of v_q on $C_c(G)$. Then this is applied to $g = \mathcal{F}f$ which is in $L_2(G)$ but not necessarily in $C_c(G)$.

4.8 **Lemma.** <u>$\mathcal{R}_2$ is spatially isomorphic to the von Neumann algebra $\mathcal{R}_3 \otimes 1$ acting in $\mathcal{H} \otimes L_2(G) \otimes L_2(G)$ where $\mathcal{R}_3$ is the von Neumann algebra in $\mathcal{H} \otimes L_2(G)$ generated by the operators</u>

$$\{x \otimes 1,\ a_s \otimes \lambda_s,\ 1 \otimes v_p \mid x \in M,\ s \in G,\ p \in \hat{G}\}.$$

Proof. We will identify $L_2(G \times G)$ with $L_2(G) \otimes L_2(G)$ by means of the map that associates to $f \otimes g$ with $f, g \in C_c(G)$ the function $(s, t) \to f(s)g(t)$ in $C_c(G \times G)$.

Then we define a unitary U in $L_2(G \times G)$ by $(Uf)(s, t) = f(st, t)$ where $f \in C_c(G \times G)$. Clearly U maps $C_c(G \times G)$ onto itself and it is easy to see that it is isometric.

Because $v_p \otimes v_p$ and $\lambda_r \otimes 1$ can easily be seen to act on $L_2(G \times G)$ as

$$((v_p \otimes v_p)f)(s, t) = \overline{\langle s, p\rangle\langle t, p\rangle} f(s, t)$$

and $((\lambda_r \otimes 1)f)(s, t) = f(r^{-1}s, t)$ where $r, s, t \in G$ and $p \in \hat{G}$ and $f \in C_c(G \times G)$, we get the following relations

$$\begin{aligned}(U^*(v_p \otimes v_p)Uf)(s, t) &= ((v_p \otimes v_p)Uf)(st^{-1}, t) = \overline{\langle st^{-1}, p\rangle}\overline{\langle t, p\rangle}(Uf)(st^{-1}, t) \\ &= \overline{\langle s, p\rangle} f(s, t) = ((v_p \otimes 1)f)(s, t)\end{aligned}$$

and

$$\begin{aligned}(U^*(\lambda_r \otimes 1)Uf)(s, t) &= ((\lambda_r \otimes 1)Uf)(st^{-1}, t) = (Uf)(r^{-1}st^{-1}, t) \\ &= f(r^{-1}s, t) = ((\lambda_r \otimes 1)f)(s, t).\end{aligned}$$

So $U^*(v_p \otimes v_p)U = v_p \otimes 1$ and $U^*(\lambda_r \otimes 1)U = \lambda_r \otimes 1$ and if we apply $(1 \otimes U^*) \cdot (1 \otimes U)$ to the operators generating $\mathcal{R}_2$ we obtain the operators

$$\{x \otimes 1 \otimes 1,\ a_s \otimes \lambda_s \otimes 1,\ 1 \otimes v_p \otimes 1 \mid x \in M,\ s \in G,\ p \in \hat{G}\}$$

proving the lemma.

We now come to the final step:

4.9 **Lemma.** $\mathcal{R}_3$ is spatially isomorphic to $M \otimes \mathcal{B}(L_2(G))$.

Proof. Let us again consider the unitary W defined by

$(W\xi)(s) = a_s \xi(s)$ for $\xi \in C_c(G, \mathcal{H})$.

We know that $\lambda(s) = 1 \otimes \lambda_s = W^*(a_s \otimes \lambda_s)W$ by proposition 2.12.

Denote by N the von Neumann algebra generated by $\{v_p \mid p \in \hat{G}\}$.

Then $N = N'$ (see appendix B). As W commutes with $1 \otimes v_p$ we have $W^*(1 \otimes N)W = 1 \otimes N$.

As $W^*(x \otimes 1)W = \pi(x)$ we also know that $W^*(M \otimes 1)W \subseteq M \otimes \mathcal{B}(L_2(G))$. In fact also $W^*(M \otimes 1)W \subseteq M \otimes N$ because $\pi(x)$ and $1 \otimes v_p$ commute. So $W^*(M \otimes N)W \subseteq M \otimes N$. Now similarly also $W(M \otimes N)W^* \subseteq M \otimes N$ so that $W^*(M \otimes N)W = M \otimes N$. Then we obtain that $W^*\mathcal{R}_3 W$ is generated by $W^*(M \otimes N)W$ and $W^*(a_s \otimes \lambda_s)W$, that is $M \otimes N$ and $1 \otimes \lambda_s$. Now because N and λ_s generate $\mathcal{B}(L_2(G))$ we get the desired result.

Combining lemmas 4.6 to 4.9 and proposition 4.5 we get the following theorem, also if α is not spatial.

4.10 Theorem. $(M \otimes_\alpha G) \otimes_{\hat{\alpha}} \hat{G}$ <u>is isomorphic to</u> $M \otimes \mathcal{B}(L_2(G))$.

We feel that the main point in this result lies in lemma 4.9 and especially in the formula

$$\{\pi(x),\ 1 \otimes v_p \mid x \in M,\ v \in \hat{G}\}'' = \{x \otimes 1,\ 1 \otimes v_p \mid x \in M,\ v \in \hat{G}\}'' .$$

In some sense this means that including the multiplication operators v_p makes it possible to get away with the 'twisting' effect of α in the definition of $\pi(x)$. This is easily illustrated in the finite case. There $\pi(x)$ is given by the diagonal matrix with elements $(\alpha_{s_1^{-1}}(x),\ \alpha_{s_2^{-1}}(x),\ \ldots,\ \alpha_{s_n^{-1}}(x))$ on the diagonal. Including the operators v_p, we include all diagonal operators with scalars on the diagonal. If we multiply $\pi(x)$ with the diagonal operator with $(1, 0, \ldots, 0)$ on the diagonal we get multiplication with $(\alpha_{s_1^{-1}}(x), 0, \ldots, 0)$. So we get also multiplication with $(x, 0, \ldots, 0)$ and we obtain all diagonal operators with entries in M.

Let us now look closer to the duality structure involved. So we started off with a covariant system (M, G, α) where M is a von Neumann algebra, G a locally compact abelian group and α a continuous action of G on M. To this triple we have associated a new one $(\hat{M}, \hat{G}, \hat{\alpha})$ where $\hat{M} = M \otimes_\alpha G$, $\hat{G}$ the dual group and $\hat{\alpha}$ the dual action. The duality would be perfect if we got the original triple back by repeating the same operation to $(\hat{M}, \hat{G}, \hat{\alpha})$. Of course the best one can ask for is to get the original system back 'up to isomorphism' whatever this may mean in the present situation. Now the dual group of $\hat{G}$ is again G, so that is no problem. We also know now what $(M \otimes_\alpha G) \otimes_{\hat{\alpha}} \hat{G}$ is, namely $M \otimes \mathcal{B}(L_2(G))$ up to isomorphism, and here real duality breaks down already, except in a special, though important case. Indeed if G is separable, then $L_2(G)$ is separable, and if M is properly infinite, then $M \otimes \mathcal{B}(L_2(G))$ can be shown to be isomorphic to M (see appendix C). So in that situation we also get the von Neumann algebra back. The question remains what happens with the bidual action $\tilde{\alpha}$, i.e. the action of G on $(M \otimes_\alpha G) \otimes_{\hat{\alpha}} \hat{G}$ dual to $\hat{\alpha}$. There we must first see how $\tilde{\alpha}$ is transformed under the isomorphism of $(M \otimes_\alpha G) \otimes_{\hat{\alpha}} \hat{G}$ with $M \otimes \mathcal{B}(L_2(G))$.

4.11 **Proposition.** <u>The bidual action $\tilde{\alpha}$ of G on $(M \otimes_\alpha G) \otimes_{\hat{\alpha}} \hat{G}$ is transformed under the isomorphism with $M \otimes \mathcal{B}(L_2(G))$ described in this section to the action θ defined in section 3.</u>

Proof. The bidual action $\tilde{\alpha}$ is dual to $\hat{\alpha}$ and implemented by the unitaries $1 \otimes 1 \otimes v_s$ where $s \in G$ and v_s is defined on $L_2(\hat{G})$ by $(v_s f)(p) = \langle s, p\rangle f(p)$. So we must see how $1 \otimes 1 \otimes v_s$ is affected by the different steps in lemmas 4.6 to 4.9.

First we have the unitary that transforms $(M \otimes_\alpha G) \otimes_{\hat{\alpha}} \hat{G}$ onto the von Neumann algebra generated by $\{\tilde{x} \otimes 1, 1 \otimes v_p \otimes \lambda_p \mid \tilde{x} \in M \otimes_\alpha G, p \in \hat{G}\}$. This unitary commutes with $1 \otimes 1 \otimes v_s$ just as W commutes with $1 \otimes v_p$, as was shown in that proof, by analogy. Of course also $W \otimes 1$ commutes with $1 \otimes 1 \otimes v_s$ and therefore the bidual action $\tilde{\alpha}$ is transformed to the action on $\mathcal{R}_1$ implemented by $1 \otimes 1 \otimes v_s$.

In lemma 4.7 we used $1 \otimes 1 \otimes \mathcal{F}$ to transform $\mathcal{R}_1$ to $\mathcal{R}_2$ and the action $\tilde{\alpha}$ will be transformed to the action on $\mathcal{R}_2$ implemented by $1 \otimes 1 \otimes \mathcal{F} v_s \mathcal{F}^*$. A similar calculation as in lemma 4.7 shows that

$$(\mathcal{F} v_s \mathcal{F}^* f)(t) = \int \overline{\langle t, p\rangle} (v_s \mathcal{F}^* f)(p) dp = \int \overline{\langle t, p\rangle} \overline{\langle s, p\rangle} (\mathcal{F}^* f)(p) dp$$

$$= \int \overline{\langle ts, p\rangle} (\mathcal{F}^* f)(p) dp = f(ts)$$

for any $f \in C_c(G)$. So $\mathfrak{F} v_s \mathfrak{F}^* = \rho_s$ as the modular function Δ is identically 1. Of course $\rho_s = \lambda_{s^{-1}}$ as G is commutative, but we will continue to use ρ_s. In lemma 4.8 we used the unitary $1 \otimes U^*$ to transform $\mathcal{R}_2$ into $\mathcal{R}_3 \otimes 1$ and the bidual action will be transformed to the action $\mathcal{R}_3 \otimes 1$ implemented by $(1 \otimes U^*)(1 \otimes 1 \otimes \rho_s)(1 \otimes U)$. Now let $f \in C_c(G \times G)$ then

$$(U^*(1 \otimes \rho_s)Uf)(r, t) = ((1 \otimes \rho_s)Uf)(rt^{-1}, t) = (Uf)(rt^{-1}, ts)$$
$$= f(rs, ts) = ((\rho_s \otimes \rho_s)f)(r, t).$$

Hence $U^*(1 \otimes \rho_s)U = \rho_s \otimes \rho_s$. In particular the bidual action transforms to the action implemented by ρ_s on $\mathcal{R}_3$. Finally in lemma 4.9 we used the unitary W again to go from $\mathcal{R}_3$ to $M \otimes \mathcal{B}(L_2(G))$ and $\tilde{\alpha}$ we will transform to the action implemented by $W^*(1 \otimes \rho_s)W$ which is $a_s \otimes \rho_s$ as for $\xi \in C_c(G, \mathcal{H})$ we have:

$$(W^*(1 \otimes \rho_s)W\xi)(t) = a_t^*(W\xi)(ts) = a_t^* a_t a_s \xi(ts) = a_s \xi(ts) = ((a_s \otimes \rho_s)\xi)(t).$$

Now if α is implemented by a_s, then θ was implemented by $a_s \otimes \rho_s$ and this completes the proof.

So we obtain that if γ is the isomorphism of $(M \otimes_\alpha G) \otimes_{\hat{\alpha}} \hat{G}$ onto $M \otimes \mathcal{B}(L_2(G))$ obtained in this section, then

$$\gamma(\tilde{\alpha}_s(\tilde{x})) = \theta_t(\gamma(\tilde{x})) \text{ for all } t \in G$$

and $\tilde{x} \in (M \otimes_\alpha G) \otimes_{\hat{\alpha}} \hat{G}$. And if we call the covariant systems (M, G, α) and (N, G, β) equivalent if there is an isomorphism τ of M onto N such that $\tau(\alpha_t(x)) = \beta_t(\tau(x))$ for any $x \in M$ then we get from the system (M, G, α) to a system equivalent to $(M \otimes \mathcal{B}(L_2(G)), G, \alpha \otimes \mathrm{ad}\,\rho)$. Unfortunately even if M is isomorphic to $M \otimes \mathcal{B}(L_2(G))$ in general one cannot hope that there is an isomorphism transforming α to $\alpha \otimes \mathrm{ad}\,\rho$. Indeed if α is trivial, $\alpha \otimes \mathrm{ad}\,\rho$ is certainly not. However one can show that there is another, weaker equivalence among triples so that the duality becomes complete [16]. We will say some more about this in the second part of our lecture notes.

To finish this section, let us prove a very much related result. So we found that the bidual action $\tilde{\alpha}$ was transformed to the action θ on $M \otimes \mathcal{B}(L_2(G))$. Now in section 3 we found that $M \otimes_\alpha G$ were precisely the fixed points in $M \otimes \mathcal{B}(L_2(G))$ under θ. It is fairly straightforward to check that $M \otimes_\alpha G$ is the image under the isomorphism γ of the subalgebra $\pi_{\hat{\alpha}}(M \otimes_\alpha G)$ of $(M \otimes_\alpha G) \otimes_{\hat{\alpha}} \hat{G}$. So it follows that $\pi_{\hat{\alpha}}(M \otimes_\alpha G)$ are the fixed points in $(M \otimes_\alpha G) \otimes_{\hat{\alpha}} \hat{G}$ for the bidual action $\tilde{\alpha}$. In fact this is a general result:

4.12 **Proposition.** $\{\tilde{x} \in M \otimes_\alpha G \,|\, \hat{\alpha}_p(\tilde{x}) = \tilde{x}$ for all $p \in \hat{G}\} = \{\pi(x) \,|\, x \in M\}$.

Proof. We obtained already that $\hat{\alpha}_p(\pi(x)) = \pi(x)$. Now if $\tilde{x} \in M \otimes_\alpha G$ and $\hat{\alpha}_p(\tilde{x}) = \tilde{x}$, then if we assume α spatial

$$\tilde{x} \in M \otimes \mathcal{B}(L_2(G)) \cap \{a_s \otimes \rho_s\}' \cap \{1 \otimes v_p\}'.$$

Now $a_s \otimes \rho_s = W^*(1 \otimes \rho_s)W$ and $1 \otimes v_p = W^*(1 \otimes v_p)W$ so that

$$\{a_s \otimes \rho_s, 1 \otimes v_p \,|\, s \in G, p \in G\}'' = W^*(1 \otimes \mathcal{B}(L_2(G))W.$$

Therefore $\tilde{x} \in M \otimes \mathcal{B}(L_2(G)) \cap W^*(\mathcal{B}(\mathcal{H}) \otimes 1)W$. So $\tilde{x} = W^*(x \otimes 1)W$ for some $x \in \mathcal{B}(\mathcal{H})$. Now if $y \in M'$ then $W^*(x \otimes 1)W$ and $y \otimes 1$ will commute. So for any $\xi \in C_c(G, \mathcal{H})$ we have $(a_s^* x a_s y - y a_s^* x a_s)\xi(s) = 0$. If $\xi(e) \neq 0$ this implies $xy - yx = 0$, so $x \in M$ and therefore $\tilde{x} = W^*(x \otimes 1)W = \pi(x)$. This completes the proof.

Remark that here we only used that $M \otimes_\alpha G$ was contained in the set of fixed points in $M \otimes \mathcal{B}(L_2(G))$ for θ. In fact in the commutative case the commutation theorem of section 3 can be obtained from proposition 4.12 and the duality theorem.

Part II · The structure of type III von Neumann algebras

1. INTRODUCTION

Let M be a von Neumann algebra and let ϕ be a faithful normal positive linear functional on M (so that in particular M has to be σ-finite). Then by the Tomita-Takesaki theory there exists a strongly continuous one-parameter group of *-automorphisms $\{\sigma_t\}$ of M characterized by the K. M. S.-condition. This says that for any pair $x, y \in M$ there is a complex-valued function F, defined, bounded and continuous on the strip $\text{Im } z \in [0, 1]$, analytic inside this strip, and with boundary values

$$F(t) = \phi(\sigma_t(x)y) \quad \text{and} \quad F(t + i) = \phi(y\sigma_t(x)). \qquad [13, 15]$$

Of course the triple $(M, \mathbf{R}, \sigma)$ is now a covariant system in the sense that $\sigma : t \to \sigma_t$ is a homomorphism of the additive group $\mathbf{R}$ into the group of *-automorphisms of M, and for each $x \in M$, the map $t \to \sigma_t(x)$ is continuous with respect to the strong topology on M. To such a triple is associated a new von Neumann algebra, called the crossed product of M by σ and is denoted by $M \otimes_\sigma \mathbf{R}$.

Because of Connes' cocycle Radon Nikodym theorem [2] it turns out that $M \otimes_\sigma \mathbf{R}$ is, up to isomorphism, independent of the faithful normal positive linear functional ϕ we started with. So to any σ-finite von Neumann algebra M is associated in a canonical way a new von Neumann algebra, let us still denote it by $M \otimes_\sigma \mathbf{R}$. Then obvious questions arise about the relationship between the properties of M and those of $M \otimes_\sigma \mathbf{R}$. As it turns out the dual action $\hat{\sigma}$, as described in part I of these notes, is important in these matters, and in fact it can be shown that also the dual action is essentially independent of the faithful normal functional ϕ we started with. All these results are covered in section 2.

In section 3 we give a proof of the existence of a semi-finite faithful normal trace τ on $M \otimes_\sigma \mathbf{R}$ with the additional property that it is relatively invariant for the dual action $\hat{\sigma}$ in the sense that $\tau(\hat{\sigma}_t(\tilde{x})) = e^{-t}\tau(\tilde{x})$ for all $t \in \mathbf{R}$ and $\tilde{x} \in M \otimes_\sigma \mathbf{R}$. Because the von Neumann algebra M here is supposed to be σ-finite, it is rather easy to avoid the theory of dual weights and left-Hilbert algebras and to give an explicit construction of such a trace. Our method however is entirely inspired by Takesaki's proof.

Finally in section 4 the cases M semi-finite and M type III are treated separately. This can be done because the crossed product $M \otimes_\sigma \mathbf{R}$ behaves nicely with respect to central decomposition in M. Because Tomita-Takesaki theory is essentially trivial for semi-finite von Neumann algebras, also $M \otimes_\sigma \mathbf{R}$ will be easily related to M. In fact one shows that $M \otimes_\sigma \mathbf{R}$ is isomorphic to $M \otimes L_\infty(\mathbf{R})$ in that case. If M is type III then it turns out that $M \otimes_\sigma \mathbf{R}$ is type II_∞. Also here our proof is different from the original one. This result together with the duality theorem for crossed products then gives a structure theorem for type III von Neumann algebras.

For results on Tomita-Takesaki theory we refer to [13, 15, 17]. We will freely use notations and results of part I of these notes. Finally numbers of lemmas and theorems will refer to results within part II except if otherwise stated.

2. CROSSED PRODUCTS WITH MODULAR ACTIONS

Let M be a von Neumann algebra on a Hilbert space $\mathcal{H}$. We will make the assumption that M is σ-finite throughout this part of these notes (except when otherwise stated). This will enable us to avoid the theory of weights and left Hilbert algebras. However it should be said here that the same (or similar) results are valid in the general case [see 16].

Because M is now σ-finite, there exist faithful normal positive linear functionals on M. Let ϕ be such a functional, then by the Tomita-Takesaki theory there is associated to it a strongly continuous one-parameter group of *-automorphisms $\{\sigma_t\}_{t\in\mathbf{R}}$ in a canonical way. Then of course σ is a continuous action of $\mathbf{R}$ on M and in this part of these notes we will be concerned with the crossed product $M \otimes_\sigma R$. Therefore let us first recall the following fundamental property [13, 15]

2.1 **Theorem.** Let ϕ be a faithful normal positive linear functional on M. Then there is a unique strongly continuous one-parameter group of *-automorphisms $\{\sigma_t\}_{t \in R}$ of M that satisfies the K.M.S.-condition with respect to ϕ, that is such that for each pair $x, y \in M$ there is a complex valued function F, defined, bounded and continuous on the strip $\mathrm{Im}\, z \in [0, 1]$, analytic in the interior of this strip, and such that

$$F(t) = \phi(\sigma_t(x)y) \text{ and } F(t+i) = \phi(y\sigma_t(x))$$

for all $t \in R$.

$\{\sigma_t\}_{t \in R}$ is called the modular automorphism group associated to ϕ. We will also call σ the modular action associated with ϕ. Occasionally we will write σ^ϕ instead of σ if there is any possible confusion.

It is well known by Tomita-Takesaki theory that ϕ is invariant for each σ_t. In fact this can be derived in the usual way by applying the above K.M.S.-condition to the pair $(x, 1)$. Then $F(t) = F(t+i) = \phi(\sigma_t(x))$ and repeating F periodically we get an entire bounded function. Therefore it is constant and in particular

$$\phi(\sigma_t(x)) = \phi(x).$$

We will consider the crossed product $M \otimes_\sigma R$ of M with the modular action σ. Because of Connes' cocycle Radon Nikodym theorem, it turns out that $M \otimes_\sigma R$ is up to isomorphism independent of the faithful normal positive linear functional ϕ we started with. For completeness let us include here also the proof of Connes' theorem [2].

2.2 **Theorem.** Let ϕ and ψ be faithful normal positive linear functionals on a von Neumann algebra M, and let σ^ϕ and σ^ψ denote the associated modular actions. Then there is a strongly continuous map $u : t \to u_t$ from R into the unitaries of M such that

(i) $\sigma_t^{\psi}(x) = u_t \sigma_t^{\phi}(x) u_t^*$

(ii) $u_{t+s} = u_t \sigma_t^{\phi}(u_s)$ $\qquad$ $t, s \in \mathbf{R}$ and $x \in M$.

In a situation like this, the actions σ^{ϕ} and σ^{ψ} are called <u>weakly equivalent</u> [16].

Proof. Let $\tilde{\mathcal{H}} = \mathcal{H} \oplus \mathcal{H}$ and denote vectors in $\tilde{\mathcal{H}}$ by columns $\binom{\xi}{\eta}$ with $\xi, \eta \in \mathcal{H}$. Then bounded operators on $\tilde{\mathcal{H}}$ are of the form $\begin{pmatrix} a & b \\ c & d \end{pmatrix}$ with $a, b, c, d \in \mathcal{B}(\mathcal{H})$. Let $\tilde{M} = \{\begin{pmatrix} a & b \\ c & d \end{pmatrix} | a, b, c, d \in M\}$, then $\tilde{M}$ is a von Neumann algebra on $\tilde{\mathcal{H}}$, in fact it is isomorphic to $M \otimes M_2$ where M_2 is the von Neumann algebra of all complex 2×2 matrices. Define a linear functional θ on $\tilde{M}$ by

$$\theta\begin{pmatrix} a & b \\ c & d \end{pmatrix} = \phi(a) + \psi(d).$$

It is not difficult to check that θ is again a faithful normal positive linear functional on $\tilde{M}$. To obtain the positivity and faithfulness one can use the fact that every positive element in $\tilde{M}$ is of the form

$$\begin{pmatrix} a & b \\ c & d \end{pmatrix}^* \begin{pmatrix} a & b \\ c & d \end{pmatrix} = \begin{pmatrix} a^* & c^* \\ b^* & d^* \end{pmatrix}\begin{pmatrix} a & b \\ c & d \end{pmatrix} = \begin{pmatrix} a^*a + c^*c & a^*b + c^*d \\ b^*a + d^*c & b^*b + d^*d \end{pmatrix}$$

with $a, b, c, d \in M$.

We will next consider the modular automorphism group σ^{θ} associated to θ on $\tilde{M}$, and see how it acts on the different matrix entries. We will use the following property: if $e \in \tilde{M}$ is such that $\theta(ex) = \theta(xe)$ for all $x \in \tilde{M}$ then $\sigma_t^{\theta}(e) = e$ for all $t \in \mathbf{R}$. To prove this one can apply the K. M. S. -condition for θ and σ^{θ} to the pair (x, e) to get a complex function F as in theorem 2.1 such that $F(t) = \theta(\sigma_t^{\theta}(x)e)$ and $F(t + i) = \theta(e\,\sigma_t^{\theta}(x))$. Now because of the property of e, again we get $F(t) = F(t + i)$ and as before that F is constant. But then $\theta(x\,\sigma_{-t}^{\theta}(e)) = \theta(\sigma_t^{\theta}(x)e) = \theta(xe)$ for all $x \in \tilde{M}$ and $t \in \mathbf{R}$ and it follows from the faithfulness of θ that $\sigma_t^{\theta}(e) = e$ for all $t \in \mathbf{R}$.

Let us now show that the element $e_{11} = \begin{pmatrix} 1 & 0 \\ 0 & 0 \end{pmatrix}$ satisfies the above property. Indeed, for any $a, b, c, d \in M$ we have

$$\theta(\begin{pmatrix} 1 & 0 \\ 0 & 0 \end{pmatrix}\begin{pmatrix} a & b \\ c & d \end{pmatrix}) = \theta(\begin{pmatrix} a & b \\ 0 & 0 \end{pmatrix}) = \phi(a)$$

and

$$\theta(\begin{pmatrix} a & b \\ c & d \end{pmatrix}\begin{pmatrix} 1 & 0 \\ 0 & 0 \end{pmatrix}) = \theta(\begin{pmatrix} a & 0 \\ c & 0 \end{pmatrix}) = \phi(a).$$

So it follows that $\sigma_t^\theta(e_{11}) = e_{11}$ for all $t \in \mathbf{R}$. Similarly we have

$$\sigma_t^\theta(e_{22}) = e_{22} \text{ for all } t \in \mathbf{R} \text{ when } e_{22} = \begin{pmatrix} 0 & 0 \\ 0 & 1 \end{pmatrix}.$$

Now trivially for any $x \in M$ we have that

$$\begin{pmatrix} x & 0 \\ 0 & 0 \end{pmatrix} = \begin{pmatrix} 1 & 0 \\ 0 & 0 \end{pmatrix}\begin{pmatrix} x & 0 \\ 0 & 0 \end{pmatrix}\begin{pmatrix} 1 & 0 \\ 0 & 0 \end{pmatrix}$$

and if we apply σ_t^θ to this relation we get

$$\sigma_t^\theta(\begin{pmatrix} x & 0 \\ 0 & 0 \end{pmatrix}) = \begin{pmatrix} 1 & 0 \\ 0 & 0 \end{pmatrix}\sigma_t^\theta(\begin{pmatrix} x & 0 \\ 0 & 0 \end{pmatrix})\begin{pmatrix} 1 & 0 \\ 0 & 0 \end{pmatrix}.$$

So for any $x \in M$ and $t \in \mathbf{R}$ there is an element $\alpha_t(x)$ in M such that

$$\sigma_t^\theta(\begin{pmatrix} x & 0 \\ 0 & 0 \end{pmatrix}) = \begin{pmatrix} \alpha_t(x) & 0 \\ 0 & 0 \end{pmatrix}.$$

It follows immediately that $\{\alpha_t\}_{t\in\mathbf{R}}$ is again a strongly continuous one-parameter group of *-automorphisms of M, and in fact, if we apply the K. M. S. -condition for θ and σ^θ on pairs of the form $(\begin{pmatrix} x & 0 \\ 0 & 0 \end{pmatrix}, \begin{pmatrix} y & 0 \\ 0 & 0 \end{pmatrix})$ with $x, y \in M$ we obtain that also α satisfies the K. M. S. -condition with respect to ϕ. Then by uniqueness it follows that $\sigma^\phi = \alpha$. Therefore we have obtained that

$$\sigma_t^\theta(\begin{pmatrix} x & 0 \\ 0 & 0 \end{pmatrix}) = \begin{pmatrix} \sigma_t^\phi(x) & 0 \\ 0 & 0 \end{pmatrix} \quad \text{for all } x \in M.$$

Similarly one obtains

$$\sigma_t^\theta(\begin{pmatrix} 0 & 0 \\ 0 & x \end{pmatrix}) = \begin{pmatrix} 0 & 0 \\ 0 & \sigma_t^\psi(x) \end{pmatrix} \quad \text{for all } x \in M.$$

Next we apply σ_t^θ to the following trivial relation

$$\begin{pmatrix} 0 & 0 \\ 1 & 0 \end{pmatrix} = \begin{pmatrix} 0 & 0 \\ 0 & 1 \end{pmatrix}\begin{pmatrix} 0 & 0 \\ 1 & 0 \end{pmatrix}\begin{pmatrix} 1 & 0 \\ 0 & 0 \end{pmatrix}$$

to get

$$\sigma_t^\theta(\begin{pmatrix} 0 & 0 \\ 1 & 0 \end{pmatrix}) = \begin{pmatrix} 0 & 0 \\ 0 & 1 \end{pmatrix}\sigma_t^\theta(\begin{pmatrix} 0 & 0 \\ 1 & 0 \end{pmatrix})\begin{pmatrix} 1 & 0 \\ 0 & 0 \end{pmatrix}.$$

It follows that for each t there is an element $u_t \in M$ such that

$$\sigma_t^\theta(\begin{pmatrix} 0 & 0 \\ 1 & 0 \end{pmatrix}) = \begin{pmatrix} 0 & 0 \\ u_t & 0 \end{pmatrix}$$

and taking adjoints

$$\sigma_t^\theta(\begin{pmatrix} 0 & 1 \\ 0 & 0 \end{pmatrix}) = \begin{pmatrix} 0 & u_t^* \\ 0 & 0 \end{pmatrix} .$$

From the relation

$$\begin{pmatrix} 0 & 0 \\ 1 & 0 \end{pmatrix}\begin{pmatrix} 0 & 1 \\ 0 & 0 \end{pmatrix} = \begin{pmatrix} 0 & 0 \\ 0 & 1 \end{pmatrix}$$

it follows that $u_t u_t^* = 1$ and similarly $u_t^* u_t = 1$. Hence u_t is unitary. The strong continuity of $t \to u_t$ follows from that of σ_t^θ at the point $\begin{pmatrix} 0 & 0 \\ 1 & 0 \end{pmatrix}$. Finally let us prove that $\{u_t\}_{t \in \mathbf{R}}$ satisfies the relations of the theorem. First apply σ_t^θ to the trivial relation

$$\begin{pmatrix} 0 & 0 \\ 0 & x \end{pmatrix} = \begin{pmatrix} 0 & 0 \\ 1 & 0 \end{pmatrix}\begin{pmatrix} x & 0 \\ 0 & 0 \end{pmatrix}\begin{pmatrix} 0 & 1 \\ 0 & 0 \end{pmatrix}$$

to get that

$$\begin{pmatrix} 0 & 0 \\ 0 & \sigma_t^\psi(x) \end{pmatrix} = \begin{pmatrix} 0 & 0 \\ u_t & 0 \end{pmatrix}\begin{pmatrix} \sigma_t^\phi(x) & 0 \\ 0 & 0 \end{pmatrix}\begin{pmatrix} 0 & u_t^* \\ 0 & 0 \end{pmatrix}$$

so that $\sigma_t^\psi(x) = u_t \sigma_t^\phi(x) u_t^*$ for all $t \in \mathbf{R}$.

Next

$$\sigma_{t+s}^\theta(\begin{pmatrix} 0 & 0 \\ 1 & 0 \end{pmatrix}) = \sigma_t^\theta(\begin{pmatrix} 0 & 0 \\ u_s & 0 \end{pmatrix}) = \sigma_t^\theta\left(\begin{pmatrix} 0 & 0 \\ 1 & 0 \end{pmatrix}\begin{pmatrix} u_s & 0 \\ 0 & 0 \end{pmatrix}\right) = \begin{pmatrix} 0 & 0 \\ u_t & 0 \end{pmatrix}\begin{pmatrix} \sigma_t^\phi(u_s) & 0 \\ 0 & 0 \end{pmatrix}$$

and it follows that $u_{t+s} = u_t \sigma_t^\phi(u_s)$.

This completes the proof.

Remark that the relation $u_{t+s} = u_t \sigma_t^\phi(u_s)$ is not really that strange. Suppose for example that there are strongly continuous one-parameter groups $\{a_t\}_{t \in \mathbf{R}}$ and $\{b_t\}_{t \in \mathbf{R}}$ of unitaries implementing actions α and β of $\mathbf{R}$ on M, and suppose that $u_t = b_t a_t^* \in M$ for all $t \in \mathbf{R}$. Then of course

$$\beta_t(x) = b_t x b_t^* = u_t a_t x a_t^* u_t^* = u_t \alpha_t(x) u_t^*$$

but also

$$\begin{aligned} u_{t+s} = b_{t+s}a^*_{t+s} &= b_t b_s a^*_s a^*_t \\ &= b_t a^*_t a_t b_s a^*_s a^*_t \\ &= u_t a_t u_s a^*_t \\ &= u_t \alpha_t(u_s). \end{aligned}$$

Let us now consider the crossed product $M \otimes_\sigma R$ where σ is now the modular action associated to the faithful normal positive linear functional ϕ. Recall that $M \otimes_\sigma R$ is a von Neumann algebra in $L_2(R, \mathcal{H})$, generated by operators $\pi(x)$ and $\lambda(t)$ defined by

$$(\pi(x)\xi)(s) = \sigma_{-s}(x)\xi(s)$$

$$(\lambda(t)\xi)(s) = \xi(s - t)$$

where $x \in M$, $t \in R$ and $\xi \in C_c(R, \mathcal{H})$, the space of continuous functions on R with values in $\mathcal{H}$ and compact support.

Using Connes' result, Takesaki now showed that the crossed product $M \otimes_\sigma R$ is independent of ϕ up to isomorphism. The proof is not very hard:

2.3 Theorem. Let ϕ and ψ be two faithful normal positive linear functionals on M, and σ^ϕ and σ^ψ the associated modular automorphism groups. Then $M \otimes_{\sigma^\phi} R$ and $M \otimes_{\sigma^\psi} R$ are isomorphic.

Proof. Remark first that $M \otimes_{\sigma^\phi} R$ and $M \otimes_{\sigma^\psi} R$ both act on the same Hilbert space. Therefore it will be sufficient to construct a unitary U on $L_2(R, \mathcal{H})$ such that $U(M \otimes_{\sigma^\phi} R)U^* = M \otimes_{\sigma^\psi} R$. Of course we use the unitaries $\{u_t\}$ in M obtained in the previous theorem.

Define U on $L_2(R, \mathcal{H})$ by $(U\xi)(s) = u_{-s}\xi(s)$ for any $\xi \in C_c(R, \mathcal{H})$. By methods similar to those used for defining $\pi(x)$, it is easy to show that $U\xi$ is again in $C_c(R, \mathcal{H})$ for $\xi \in C_c(R, \mathcal{H})$ and that U extends to an isometry on $L_2(R, \mathcal{H})$. Moreover $(U^*\xi)(s) = u^*_{-s}\xi(s)$ for $\xi \in C_c(R, \mathcal{H})$ and U will be unitary as the u_s are unitaries. (Remark that in general u_{-s} will not equal u_s^* as $\{u_s\}$ need not be a one-parameter group of unitaries.)

Now denote by π_ϕ and π_ψ the associated representations of M on $L_2(\mathbf{R}, \mathcal{H})$. Then, with $\xi \in C_c(\mathbf{R}, \mathcal{H})$ and $s \in \mathbf{R}$,

$$
\begin{aligned}
(U\pi_\phi(x)U^*\xi)(s) &= u_{-s}(\pi_\phi(x)U^*\xi)(s) \\
&= u_{-s}\sigma^\phi_{-s}(x)(U^*\xi)(s) \\
&= u_{-s}\sigma^\phi_{-s}(x)u^*_{-s}\xi(s) \\
&= \sigma^\psi_{-s}(x)\xi(s) \\
&= (\pi_\psi(x)\xi)(s)
\end{aligned}
$$

and

$$
\begin{aligned}
(U\lambda(t)U^*\xi)(s) &= u_{-s}(\lambda(t)U^*\xi)(s) \\
&= u_{-s}(U^*\xi)(s - t) \\
&= u_{-s}u^*_{t-s}\xi(s - t) \\
&= u_{-s}(u_{-s}\sigma^\phi_{-s}(u_t))^*\xi(s - t) \\
&= u_{-s}\sigma^\phi_{-s}(u_t^*)u^*_{-s}\xi(s - t) \\
&= \sigma^\psi_{-s}(u_t^*)\xi(s - t) \\
&= \sigma^\psi_{-s}(u_t^*)(\lambda(t)\xi)(s) \\
&= (\pi_\psi(u_t^*)\lambda(t)\xi)(s).
\end{aligned}
$$

So $U\pi_\phi(x)U^* = \pi_\psi(x)$ for all $x \in M$ and $U\lambda(t)U^* = \pi_\psi(u_t^*)\lambda(t)$. It follows already that $U(M \otimes_{\sigma^\phi} \mathbf{R})U^* \subseteq M \otimes_{\sigma^\psi} \mathbf{R}$. Now we also get $U^*\pi_\psi(x)U = \pi_\phi(x)$ and $\lambda(t) = U^*\pi_\psi(u_t^*)U \cdot U^*\lambda(t)U$ so that $U^*\lambda(t)U = \pi_\phi(u_t)\lambda(t)$ and it also follows that $U^*(M \otimes_{\sigma^\psi} \mathbf{R})U \subseteq M \otimes_{\sigma^\phi} \mathbf{R}$. Therefore we have equality and the proof is complete.

So to any σ-finite von Neumann algebra we have associated a new von Neumann algebra in a canonical way. Then the obvious questions arise about the relation of the properties of both von Neumann algebras. Very much related to those questions will be the dual action. Indeed because $\mathbf{R}$ is a commutative group we can define a dual action $\hat{\sigma}$ on $M \otimes_\sigma \mathbf{R}$ of the dual group of $\mathbf{R}$ which we of course identify with $\mathbf{R}$ itself as usual by putting $\langle s, t\rangle = e^{its}$. Recall that the dual action was defined by means of a unitary representation v of the dual group $\hat{G}$ on $L_2(G)$. This was defined by $(v_p f)(s) = \overline{\langle s, p\rangle} f(s)$ for $f \in C_c(G)$. So here we have $(v_t f)(s) = e^{-its} f(s)$ for any $f \in C_c(\mathbf{R})$. Now we also had identified $\mathcal{H} \otimes L_2(\mathbf{R})$ and $L_2(\mathbf{R}, \mathcal{H})$ in such a way that for any $\xi \in \mathcal{H}$ and $f \in C_c(\mathbf{R})$ the vector $\xi \otimes f$ was considered as a function in $C_c(\mathbf{R}, \mathcal{H})$ with values $(\xi \otimes f)(s) = f(s)\xi$ for $s \in \mathbf{R}$. Then the dual action on $M \otimes_\sigma \mathbf{R}$ was defined by

$$\hat{\sigma}_t(\tilde{x}) = (1 \otimes v_t)\tilde{x}(1 \otimes v_t^*) \quad \text{with } \tilde{x} \in M \otimes_\sigma \mathbf{R}.$$

We have seen that up to isomorphism, the crossed product $M \otimes_\sigma \mathbf{R}$ was independent of ϕ. In the following sense this is also true for the dual action.

2.4 Proposition. <u>Let ϕ and ψ be two faithful normal positive linear functionals on M, and as before let σ^ϕ and σ^ψ be the associated modular actions. Then if γ is the isomorphism of $M \otimes_{\sigma^\phi} \mathbf{R}$ onto $M \otimes_{\sigma^\psi} \mathbf{R}$ obtained in theorem 2.3 we will have</u>

$$\hat{\sigma}_t^\psi(\gamma(\tilde{x})) = \gamma(\hat{\sigma}_t^\phi(\tilde{x}))$$

<u>for $\tilde{x} \in M \otimes_{\sigma^\phi} \mathbf{R}$, where $\hat{\sigma}^\phi$ and $\hat{\sigma}^\psi$ are the actions dual to σ^ϕ and σ^ψ respectively.</u>

Proof. Let $\{u_t\}$ be the unitaries of theorem 2.2 and let U be defined as in theorem 2.3 by $(U\xi)(s) = u_{-s}\xi(s)$. Then clearly $U(1 \otimes v_t) = (1 \otimes v_t)U$. Then

$$\hat{\sigma}_t^{\psi}(\gamma(\tilde{x})) = (1 \otimes v_t) U \tilde{x} U^*(1 \otimes v_t^*)$$
$$= U(1 \otimes v_t)\tilde{x}(1 \otimes v_t^*)U^*$$
$$= U \hat{\sigma}_t^{\phi}(\tilde{x})U^*$$
$$= \gamma(\hat{\sigma}_t^{\phi}(\tilde{x})).$$

So to a σ-finite von Neumann algebra M there is a now associated in a canonical way a new von Neumann algebra, together with a strongly continuous one-parameter group of *-automorphisms. It makes sense to denote this new von Neumann algebra simply as $M \otimes_\sigma \mathbf{R}$ and the one-parameter group as $\hat{\sigma}$. The couple $(M \otimes_\sigma \mathbf{R}, \hat{\sigma})$ is defined up to isomorphism (in the above sense) by M. In the next two sections we will see what properties can be obtained about $M \otimes_\sigma \mathbf{R}$ and the action $\hat{\sigma}$ from properties of M. Let us finish this section here by remarking that M can be recovered from the pair $(M \otimes_\sigma \mathbf{R}, \hat{\sigma})$. Indeed by proposition 4.12 of part I we know that M is isomorphic to the fixed points in $M \otimes_\sigma \mathbf{R}$ under $\hat{\sigma}$. So

$$M \cong \{\tilde{x} \in M \otimes_\sigma \mathbf{R} \,|\, \hat{\sigma}_t(\tilde{x}) = \tilde{x}, \quad \forall t \in \mathbf{R}\}.$$

If M is properly infinite there is another way of recovering M from $M \otimes_\sigma \mathbf{R}$ and σ. Indeed, as we have seen $(M \otimes_\sigma \mathbf{R}) \otimes_{\hat{\sigma}} \mathbf{R} \cong M \otimes \mathcal{B}(L_2(\mathbf{R})) \cong M$ so that M is the crossed product of $M \otimes_\sigma \mathbf{R}$ by the action $\hat{\sigma}$ of $\mathbf{R}$. We also mention here that the actions σ on M and $\sigma \otimes \mathrm{ad}\, \rho$ on $M \otimes \mathcal{B}(L_2(\mathbf{R}))$ are weakly equivalent in the sense of [16, theorem 2.2]. This makes the duality in section 4 of part I in some sense more complete.

3. THE SEMI-FINITENESS OF $M \otimes_\sigma \mathbf{R}$

In this section we give a proof of the fact that $M \otimes_\sigma \mathbf{R}$ is semi-finite. In fact there always exists a faithful normal semi-finite trace τ on $M \otimes_\sigma \mathbf{R}$ such that $\tau(\hat{\sigma}_t(\tilde{x})) = e^{-t}\tau(\tilde{x})$ for all $\tilde{x} \in M \otimes_\sigma \mathbf{R}$. Such a trace is called relatively invariant for $\hat{\sigma}$. We give an explicit construction of such a trace, and we should mention here that we were very much inspired by Takesaki's approach [16].

So assume that ϕ is a faithful normal positive linear functional and that σ is the associated modular action. We may assume the existence of a separating and cyclic vector $\omega \in \mathcal{H}$ such that $\phi(x) = \langle x\omega, \omega\rangle$ for all $x \in M$, and of a strongly continuous one-parameter group $\{a_t\}_{t \in \mathbf{R}}$ of unitaries such that $\sigma_t(x) = a_t x a_t^*$ and $a_t\omega = \omega$ for all $t \in \mathbf{R}$. This follows easily from the G. N. S. -construction associated with ϕ because ϕ is invariant for σ. Of course one could also consider Δ^{it} for a_t where Δ is the modular operator associated to ω.

Before we start let us first agree on a number of notations.

3.1 **Notations.** Denote by $\mathcal{F}$ the Fourier transform in $L_2(\mathbf{R})$; so

$$(\mathcal{F}f)(t) = \frac{1}{\sqrt{2\pi}} \int e^{-its} f(s)ds \quad \text{for all } f \in C_c(\mathbf{R})$$

where now ds is the Lebesgue measure on $\mathbf{R}$. Then $\mathcal{F}$ is a unitary in $L_2(\mathbf{R})$ and we have $\mathcal{F}^*\lambda_s\mathcal{F} = v_{-s}$ for all $s \in \mathbf{R}$ by a straightforward calculation. Let $f \in L_\infty(\mathbf{R})$ and let as before denote by m_f the multiplication operator in $L_2(\mathbf{R})$ by the function f. Also denote $\lambda_f = \mathcal{F}^* m_f \mathcal{F}$ and $\lambda(f) = 1 \otimes \lambda_f$. Because m_f belongs to the von Neumann algebra generated by the operators $\{v_s,\ s \in \mathbf{R}\}$, also λ_f will belong to the one generated by $\{\lambda_s,\ s \in \mathbf{R}\}$ and therefore $\lambda(f) \in M \otimes_\sigma \mathbf{R}$ [appendix B].

Finally let K be any compact subset of $\mathbf{R}$, define $f_K(s) = \chi_K(s)\exp\frac{s}{2}$ and $\xi_K = \omega \otimes \mathcal{F}^* f_K$. We put $\tau_K(\tilde{x}) = \langle \tilde{x}\xi_K, \xi_K\rangle$ for any $\tilde{x} \in M \otimes_\sigma \mathbf{R}$. We will show that τ_K increases with K, and increases to the desired trace. We need a number of lemmas.

3.2 **Lemma.** <u>For any pair</u> $f, g \in L_2(\mathbf{R})$ <u>and</u> $x \in M$ <u>we have</u>

$$\langle \pi(x)\omega \otimes f, \omega \otimes g\rangle = \phi(x)\langle f, g\rangle.$$

Proof.
$$\begin{aligned}\langle \pi(x)\omega \otimes f, \omega \otimes g\rangle &= \int \langle (\pi(x)\omega \otimes f)(s), (\omega \otimes g)(s)\rangle ds \\ &= \int \langle \sigma_{-s}(x)f(s)\omega, g(s)\omega\rangle ds \\ &= \int \phi(\sigma_{-s}(x))f(s)\overline{g(s)}ds\end{aligned}$$

$$= \phi(x) \int f(s)\overline{g(s)}ds$$

$$= \phi(x)\langle f, g\rangle.$$

3.3 **Lemma.** If $f \in C_c(\mathbf{R})$ and has support in the compact set K, then

$$\tau_K(\pi(x)\lambda(s)\lambda(f)) = \sqrt{2\pi}\,\hat{f}(i + s)\phi(x)$$

where

$$\hat{f}(z) = \frac{1}{\sqrt{2\pi}} \int e^{-izt} f(t)dt$$

for all complex z.

Proof.

$$\begin{aligned}\tau_K(\pi(x)\lambda(s)\lambda(f)) &= \langle \pi(x)\lambda(s)\lambda(f)\omega \otimes \mathfrak{F}^*f_K, \omega \otimes \mathfrak{F}^*f_K\rangle \\ &= \langle \pi(x)\omega \otimes \lambda_s\lambda_f\mathfrak{F}^*f_K, \omega \otimes \mathfrak{F}^*f_K\rangle \\ &= \phi(x)\langle \lambda_s\lambda_f\mathfrak{F}^*f_K, \mathfrak{F}^*f_K\rangle \\ &= \phi(x)\langle v_s m_f f_K, f_K\rangle \\ &= \phi(x) \int e^{-its} f(t)\chi_K(t)e^t dt \\ &= \phi(x) \int e^{-its} e^t f(t)dt \\ &= \phi(x)\sqrt{2\pi}\,\hat{f}(i + s)\end{aligned}$$

where we have used lemma 3.2 and the fact that f has support in K.

3.4 **Lemma.** Let f and g be continuous functions with support in a compact set K and such that the Fourier transforms are again L_1. Then

$$\tau_K(\pi(x)\lambda(f)\lambda(y)\lambda(g)) = \tau_K(\pi(y)\lambda(g)\pi(x)\lambda(f)).$$

Proof. Because the Fourier transform $\hat{f}$ of f is again L_1 we have that $f(t) = \frac{1}{\sqrt{2\pi}} \int e^{its}\hat{f}(s)ds$. So, using methods as in part 1, section 3, we also have $m_f = \frac{1}{\sqrt{2\pi}} \int \hat{f}(s)v_{-s}ds$ in the σ-weak topology, and applying $\mathfrak{F}^* \cdot \mathfrak{F}$ that

$$\lambda_f = \frac{1}{\sqrt{2\pi}} \int \hat{f}(s)\lambda_{-s}ds\ .$$

So $\tau_K(\pi(x)\lambda(f)\pi(y)\lambda(g)) = \frac{1}{\sqrt{2\pi}} \int \hat{f}(s)\tau_K(\pi(x)\lambda(-s)\pi(y)\lambda(g))ds$

$$= \frac{1}{\sqrt{2\pi}} \int \hat{f}(s)\tau_K(\pi(x\sigma_{-s}(y))\lambda(-s)\lambda(g))ds$$

$$= \int \hat{f}(s)\hat{g}(i - s)\phi(x\sigma_{-s}(y))ds$$

by lemma 3.3.

Now by the K. M. S. property there is a complex valued function F, defined, bounded and continuous on the strip $\operatorname{Im} z \in [0, 1]$, analytic inside this strip, and with boundary values $F(s) = \phi(\sigma_s(x)y) = \phi(x\sigma_{-s}(y))$ and $F(s + i) = \phi(y\sigma_s(x))$. Now because f and g have compact support, $\hat{f}$ and $\hat{g}$ will be analytic everywhere and we apply Cauchy's formula to the function $\hat{f}(z)\hat{g}(i - z)F(z)$. We integrate along the curve

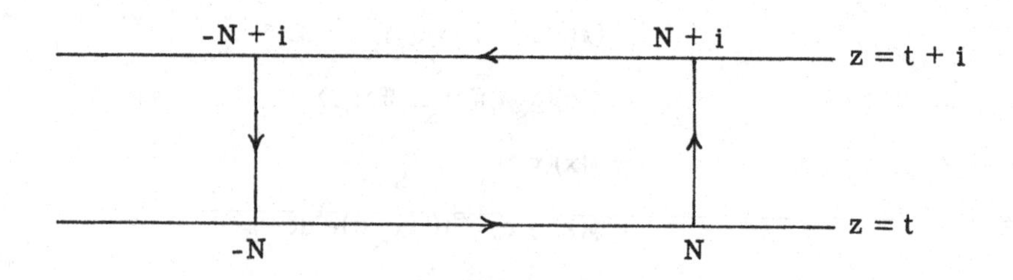

Let us show that the integrals over the vertical lines tend to zero when $N \to \infty$. First remark that for any $a \in [0, 1]$ we have

$$\hat{f}(N + ia) = \frac{1}{\sqrt{2\pi}} \int e^{-iNt}e^{at}f(t)dt$$

so that $\lim_{N\to\infty} \hat{f}(N + ia) = 0$ because the Fourier transform of any L_1-function, and therefore of any $C_c(\mathbf{R})$ function tends to zero at infinity. Also

$$|\hat{f}(N + ia)| \leq \frac{1}{\sqrt{2\pi}} \left(\int |f(t)|dt\right) \sup_{t\in K} (\max(1, e^t)).$$

Similarly $\hat{g}(i - N - ia) = \hat{g}(-N + i(1 - a))$ is uniformly bounded and tends to zero for $N \to \infty$. Then by the dominated convergence theorem we have

$$\lim_{N\to\infty} \int_0^1 \hat{f}(N + ia)\hat{g}(-N + i(1 - a))F(N + ia)da = 0.$$

Similarly for the other vertical line and it follows that

$$\int_{-\infty}^{+\infty}\hat{f}(s)\hat{g}(i-s)F(s)ds = \int_{-\infty}^{+\infty}\hat{f}(s+i)\hat{g}(-s)F(s+i)ds = \int_{-\infty}^{+\infty}\hat{g}(s)\hat{f}(i-s)F(i-s)ds.$$

Now $F(s) = \phi(x\sigma_{-s}(y))$ and $F(i - s) = \phi(y\sigma_{-s}(x))$ so that

$$\tau_K(\pi(x)\lambda(f)\pi(y)\lambda(g)) = \int \hat{f}(s)\hat{g}(i - s)\phi(x\sigma_{-s}(y))ds$$
$$= \int \hat{g}(s)\hat{f}(i - s)\phi(y\sigma_{-s}(x))ds$$

and therefore by symmetry we get

$$\tau_K(\pi(x)\lambda(f)\pi(y)\lambda(g)) = \tau_K(\pi(y)\lambda(g)\pi(x)\lambda(f)).$$

3.4 **Lemma.** <u>If K is a compact set in $\mathbf{R}$ and $p_K = \lambda(\chi_K)$ then τ_K is a trace on $p_K(M \otimes_\sigma \mathbf{R})p_K$.</u>

Proof. Let K_1 be a compact set whose interior contains K, choose a C^∞-function f_0 with support in K_1 and such that $f_0(s) = 1$ for $s \in K$. Then for any $s, t \in \mathbf{R}$ we will still have that $f = v_s f_0$ and $g = v_t f_0$ are C^∞-functions with support in K_1 and we can apply the previous lemma to f and g. Because $\lambda_f = \mathfrak{F}^* m_f \mathfrak{F}$ and $m_f = v_s m_{f_0}$ we get $\lambda(f) = \lambda(s)\lambda(f_0)$ and so

$$\tau_{K_1}(\pi(x)\lambda(s)\lambda(f_0)\pi(y)\lambda(t)\lambda(f_0)) = \tau_{K_1}(\pi(y)\lambda(t)\lambda(f_0)\pi(x)\lambda(s)\lambda(f_0)).$$

By linearity and continuity and the fact that operators of the form $\pi(x)\lambda(s)$ span a dense subalgebra of $M \otimes_\sigma \mathbf{R}$ we get

$$\tau_{K_1}(\tilde{x}\lambda(f_0)\tilde{y}\lambda(f_0)) = \tau_{K_1}(\tilde{y}\lambda(f_0)\tilde{x}\lambda(f_0))$$

for all $\tilde{x}, \tilde{y} \in M \otimes_\sigma \mathbf{R}$. Then replace $\tilde{x}$ and $\tilde{y}$ by $p_K\tilde{x}p_K$ and $p_K\tilde{y}p_K$, as $p_K\lambda(f_0) = \lambda(\chi_K f_0) = \lambda(\chi_K) = p_K$ we obtain

$$\tau_{K_1}(p_K\tilde{x}p_K\tilde{y}p_K) = \tau_{K_1}(p_K\tilde{y}p_K\tilde{x}p_K).$$

Finally

$$p_K \xi_{K_1} = \omega \otimes \lambda_{\chi_K} \mathfrak{F}^* \xi_{K_1} = \omega \otimes \mathfrak{F}^* m_{\chi_K} f_{K_1}$$

$$= \omega \otimes \mathfrak{F}^* f_K \quad \text{as} \quad f_K(s) = \chi_K(s) e^{s/2}$$

$$= \xi_K$$

and τ_{K_1} restricted to $p_K(M \otimes_\sigma R)p_K$ equals τ_K. This completes the proof.

We finally come to the main theorem

3.5 **Theorem.** Define $\tau = \sup_K \tau_K$ on $(M \otimes_\sigma R)^+$, then τ is a faithful normal semi-finite trace on $M \otimes_\sigma R$ such that $\tau(\hat{\sigma}_t(\tilde{x})) = e^{-t}\tau(\tilde{x})$ for all $t \in R$ and $\tilde{x} \in (M \otimes_\sigma R)^+$.

Proof. Suppose that K and K_1 are compact in R and that $K \subseteq K_1$. Then

$$\tau_{K_1}(\pi(x)\lambda(s)) = \phi(x) \int_{K_1} e^{-its} e^t dt$$

as in lemma 3.3 and therefore it is easy to see that

$$\tau_{K_1}(\pi(x)\lambda(s)) = \tau_K(\pi(x)\lambda(s)) + \langle \pi(x)\lambda(s)(\xi_{K_1} - \xi_K), (\xi_{K_1} - \xi_K)\rangle$$

so that $\tau_{K_1} \geq \tau_K$. It follows that $\tau = \sup_K \tau_K$ defines a map from $(M \otimes_\sigma R)^+$ to $[0, \infty]$ such that $\tau(\tilde{x} + \tilde{y}) = \tau(\tilde{x}) + \tau(\tilde{y})$ and $\tau(\alpha\tilde{x}) = \alpha\tau(\tilde{x})$ for all $\tilde{x}, \tilde{y} \in M \otimes_\sigma R$ and $\alpha > 0$.

Let us prove that τ is a trace. Take $\tilde{x} \in M \otimes_\sigma R$ and K, K_1 compact with $K \subseteq K_1$, then

$$\tau_{K_1}(\tilde{x}^* p_K \tilde{x}) = \tau_{K_1}(p_{K_1} \tilde{x}^* p_K \tilde{x} p_{K_1})$$

$$= \tau_{K_1}(p_K \tilde{x} p_{K_1} \tilde{x}^* p_K) \qquad \text{(lemma 3.4)}$$

$$= \tau_K(\tilde{x} p_{K_1} \tilde{x}^*).$$

If K_1 tends to R then

$$\tau(\tilde{x}^* p_K \tilde{x}) = \tau_K(\tilde{x}\tilde{x}^*) \text{ as } p_{K_1} \nearrow 1.$$

Now τ is normal as the supremum of normal functionals, so again by letting $\mathbf{K}$ tend to $\mathbf{R}$ we get

$$\tau(\tilde{x}^* \tilde{x}) = \tau(\tilde{x}\, \tilde{x}^*).$$

Furthermore τ is semi-finite as it is finite on a dense subalgebra, namely $\bigcup_{\mathbf{K}} p_{\mathbf{K}}(M \otimes_\sigma \mathbf{R})p_{\mathbf{K}}$.

To show that τ is faithful, it is sufficient to show that $\{\xi_{\mathbf{K}}\}_{\mathbf{K}\text{ compact}}$ is cyclic for $(M \otimes_\sigma \mathbf{R})'$. For any $x' \in M'$ we know that $x' \otimes 1 \in (M \otimes_\sigma \mathbf{R})'$ and $(x' \otimes 1)\xi_{\mathbf{K}} = x'\omega \otimes \mathcal{F}^* f_{\mathbf{K}}$ and those vectors span a dense subspace of $\mathcal{H} \otimes L_2(\mathbf{R})$ as ω is cyclic for M' and as $\mathbf{K}$ runs over all compact subsets of $\mathbf{R}$.

To complete the proof we show that τ is relatively invariant. So let $\mathbf{K}$ be compact and $t \in \mathbf{R}$, then $v_t^* \mathcal{F}^* f_{\mathbf{K}} = \mathcal{F}^* \mathcal{F} v_t^* \mathcal{F}^* f_{\mathbf{K}}$ and a straightforward calculation shows that $\mathcal{F} v_t^* \mathcal{F}^* = \lambda_t$. Now

$$\begin{aligned}(\lambda_t f_{\mathbf{K}})(s) = f_{\mathbf{K}}(s - t) &= \chi_{\mathbf{K}}(s - t)\exp(s - t)/2 \\ &= \chi_{\mathbf{K}+t}(s)\exp\tfrac{s}{2} \cdot \exp(-\tfrac{t}{2}) \\ &= e^{-t/2} \cdot f_{\mathbf{K}+t}(s).\end{aligned}$$

Therefore, for any $\tilde{x} \in (M \otimes_\sigma \mathbf{R})^+$, we have

$$\begin{aligned}\tau_{\mathbf{K}}(\hat{\sigma}_t(\tilde{x})) &= \tau_{\mathbf{K}}((1 \otimes v_t)\tilde{x}(1 \otimes v_t^*)) \\ &= \langle \tilde{x}\, \omega \otimes v_t^* \mathcal{F}^* f_{\mathbf{K}},\ \omega \otimes v_t^* \mathcal{F}^* f_{\mathbf{K}} \rangle \\ &= \langle \tilde{x}\, e^{-t/2}\omega \otimes \mathcal{F}^* f_{\mathbf{K}+t},\ e^{-t/2}\omega \otimes \mathcal{F}^* f_{\mathbf{K}+t} \rangle \\ &= e^{-t}\tau_{\mathbf{K}+t}(\tilde{x}).\end{aligned}$$

Then taking supremum over $\mathbf{K}$ we get

$$\tau(\hat{\sigma}_t(\tilde{x})) = e^{-t}\tau(\tilde{x}).$$

This completes the proof.

We want to finish this section be some remarks. First, if we look at the formula of lemma 3.3, we see that ϕ can easily be reconstructed from τ. Indeed for any function $f \in C_c(\mathbf{R})$ we will have that

$\tau(\lambda(f)) < \infty$ and that

$$\phi(x) = \frac{1}{\sqrt{2\pi}} \frac{1}{\hat{f}(i)} \quad \tau(\pi(x)\lambda(f))$$

provided of course $\hat{f}(i) \neq 0$. Remark also that the formula in that lemma is very much related to lemma 5.19 of [16].

Finally, if the von Neumann algebra is not σ-finite the preceding construction can be modified to work in the case where ϕ is a strictly semi-finite weight, although it would be more complicated if ϕ is just any faithful normal semi-finite weight.

4. THE STRUCTURE OF TYPE III VON NEUMANN ALGEBRAS

In this section we want to say something about the type of the crossed product $M \otimes_\sigma R$ given the type of M. Therefore let us first see how central decomposition in M is reflected in the crossed product.

4.1 **Proposition.** *If p is a central projection in M, then $p \otimes 1$ is a central projection in $M \otimes_\sigma R$. Moreover Mp is left invariant by the action σ and if $\bar{\sigma}$ denotes the restriction of σ to Mp then*

$$(M \otimes_\sigma R)(p \otimes 1) = Mp \otimes_{\bar{\sigma}} R.$$

Proof. If p is in the centre of M it is well known that $\sigma_t(p) = p$ for any modular action. In fact if σ is the modular action associated to the faithful normal positive linear functional ϕ, then of course $\phi(px) = \phi(xp)$ for all $x \in M$ and as in the proof of theorem 2.3 it would follow that $\sigma_t(p) = p$. Then from a trivial calculation it follows that $\pi(p) = p \otimes 1$. Now because $p \in M'$ it follows that also $p \otimes 1 \in (M \otimes_\sigma R)'$ so that $p \otimes 1$ is in the centre of $M \otimes_\sigma R$.

Because $\sigma_t(p) = p$ for all t it also follows that σ leaves the von Neumann algebra Mp invariant.

Finally $(M \otimes_\sigma R)(p \otimes 1)$ is the von Neumann algebra on $p\mathcal{H} \otimes L_2(R)$ generated by $\pi(x)(p \otimes 1) = \pi(xp)$ and $\lambda(s)(p \otimes 1) = p \otimes \lambda_s$. $\pi(xp)$ restricted to the space $L_2(R, p\mathcal{H})$ is clearly $\pi_{\bar{\sigma}}(xp)$ where $\pi_{\bar{\sigma}}$ is the representation associated to $\bar{\sigma}$ and $p \otimes \lambda_s$ restricted to

$L_2(R, p\mathcal{H})$ is $1 \otimes \lambda_s$.

Then the result follows.

We will now consider the cases M semi-finite and M type III separately. This makes sense because of the previous proposition.

If M is semi-finite it is well known that the modular automorphisms are inner [15]. In fact this could be deduced from Connes' cocycle Radon-Nikodym theorem for weights and the fact that the modular action associated to the trace is the trivial one. In an appendix we present a fairly easy proof of this fact for σ-finite von Neumann algebras (appendix D).

We get the following result for semi-finite von Neumann algebras:

4.2 Proposition. If M is semi-finite there exists an isomorphism of $M \otimes_\sigma R$ onto $M \otimes L_\infty(R)$ that transforms the dual action to $1 \otimes \lambda$ where λ acts on $L_\infty(R)$ by $(\lambda_t f)(s) = f(s - t)$ as before.

Proof. So assume that $t \to a_t$ is a continuous unitary representation of R on $\mathcal{H}$ with $a_t \in M$ and $\sigma_t(x) = a_t x a_t^*$ for all $t \in R$ and $x \in M$. Then as we have seen in part 1, there is an isomorphism of $M \otimes_\sigma R$ onto the algebra generated by $\{x \otimes 1, a_t \otimes \lambda_t\}$ and the dual action is transformed to the action implemented by $1 \otimes v_s$. Now because $a_t \in M$ we have $a_t \otimes \lambda_t \in \{x \otimes 1, 1 \otimes \lambda_t\}''$, and also $1 \otimes \lambda_t = (a_t^* \otimes 1)(a_t \otimes \lambda_t) \in \{x \otimes 1, a_t \otimes \lambda_t\}''$. Therefore $(M \otimes_\sigma R) = \{x \otimes 1, 1 \otimes \lambda_t\}''$. Then by applying the Fourier transform we get an isomorphism of $(M \otimes_\sigma R)$ onto $M \otimes L_\infty(R)$ because $\{v_t\}'' = L_\infty(R)$ and under this isomorphism the dual action is transformed to the action implemented by $1 \otimes \lambda_t$.

So we see that the whole theory is not interesting for semi-finite von Neumann algebras as was to be expected because Tomita-Takesaki theory is more or less trivial in this case. The type III case is much more interesting, as we will show it turns out that in that case $M \otimes_\sigma R$ is of type II_∞.

Let us first show that if $M \otimes_\sigma R$ is type I, then M is semi-finite. Then the result on type III will follow easily because the crossed product behaves nicely w.r.t. central decomposition. We have divided the argument in a number of steps.

4.3 **Lemma.** Let τ be the relatively invariant trace on $M \otimes_\sigma R$ constructed in section 3. Then there is a projection p with the properties that $\tau(p) < \infty$ and $\hat{\sigma}_t(p)$ is increasing when t decreases and

$$\lim_{t \to +\infty} \hat{\sigma}_t(p) = 0 \qquad \lim_{t \to -\infty} \hat{\sigma}_t(p) = 1.$$

Proof. With the notations of the previous section, let $p = \lambda(\chi_{(-\infty, 0]})$. For any compact set K in R we have

$$\begin{aligned}
\tau_K(p) &= \langle p\, \omega \otimes \xi_K,\ \omega \otimes \xi_K \rangle \\
&= \langle \omega, \omega \rangle \langle \mathfrak{F}^* m_{\chi_{(-\infty, 0]}} \mathfrak{F}\mathfrak{F}^* f_K,\ \mathfrak{F}^* f_K \rangle \\
&= \langle \omega, \omega \rangle \langle m_{\chi_{(-\infty, 0]}} f_K,\ f_K \rangle \\
&= \langle \omega, \omega \rangle \int_{-\infty}^{0} \chi_K(t) e^t dt.
\end{aligned}$$

So $\tau(p) = \langle \omega, \omega \rangle \int_{-\infty}^{0} e^t dt < \infty$.

Furthermore
$$\begin{aligned}
\hat{\sigma}_t(p) &= 1 \otimes v_t \lambda_{\chi_{(-\infty, 0]}} v_t^* \\
&= 1 \otimes v_t \mathfrak{F}^* m_{\chi_{(-\infty, 0]}} \mathfrak{F} v_t^* \\
&= 1 \otimes \mathfrak{F}^* \lambda_{-t} m_{\chi_{(-\infty, 0]}} \lambda_t \mathfrak{F} \\
&= 1 \otimes \mathfrak{F}^* m_{\chi_{(-\infty, -t]}} \mathfrak{F} \\
&= \lambda(\chi_{(-\infty, -t]}).
\end{aligned}$$

Then the result follows easily.

Now let $q = c(p)$, the central support of p in $M \otimes_\sigma R$. Clearly $\hat{\sigma}_t(q) = c(\hat{\sigma}_t(p))$ and because $\hat{\sigma}_t(p)$ increases when t decreases, the same will be true for $\hat{\sigma}_t(q)$. Because $\hat{\sigma}_t(q) \geq \hat{\sigma}_t(p)$ we will still have that

$$\lim_{t \to -\infty} \hat{\sigma}_t(q) = 1$$

but in general we can not say anything about $\lim \hat{\sigma}_t(q)$ when $t \to +\infty$. We can only say that the limit exists. Here we will use the following lemma.

4.4 Lemma. *If e is an abelian projection in a von Neumann algebra then* $e\,c(f) \precsim f$ *for any projection* f *where* $c(f)$ *is the central support of* f *and* $\precsim$ *stands for the ordering in the projections relative to the von Neumann algebra.*

Proof. $e_1 = e\,c(f)$ is again an abelian projection, majorised by $c(f)$. Then it follows that $c(e_1) \le c(f)$ and since e_1 is abelian by [6, p. 239] it follows that $e_1 \precsim f$.

4.5 Lemma. *Suppose now that* $M \otimes_\sigma R$ *is type I. If* $q = c(p)$, *the central support of the projection* p *of lemma 4.3, then* $\hat{\sigma}_t(q)$ *is increasing when* t *decreases and*

$$\lim_{t \to +\infty} \hat{\sigma}_t(q) = 0 \qquad\qquad \lim_{t \to -\infty} \hat{\sigma}_t(q) = 1.$$

Proof. As $\hat{\sigma}_t(q) = c(\hat{\sigma}_t(p))$ we remarked already that also $\hat{\sigma}_t(q)$ will increase when t decreases. And as $\hat{\sigma}_t(q) \ge \hat{\sigma}_t(p)$ we also have that $\lim \hat{\sigma}_t(q) = 1$ when $t \to -\infty$. Now let $q_0 = \lim \hat{\sigma}_t(q)$ for $t \to +\infty$. For any abelian projection we have that $e\hat{\sigma}_t(q) = e\,c(\hat{\sigma}_t(p)) \precsim \hat{\sigma}_t(p)$ by the previous lemma so that $\tau(e\,\hat{\sigma}_t(q)) \le \tau(\hat{\sigma}_t(p)) = e^{-t}\tau(p)$. As $\tau(p) < \infty$ we will have by the normality of the trace that $\tau(e\,\hat{\sigma}_t(q)) \to \tau(e\,q_0)$ when $t \to +\infty$ and that $\tau(e\,q_0) = 0$. Then because τ is faithful we get $e\,q_0 = 0$ for all abelian projections e. Because q_0 is in the centre and $M \otimes_\sigma R$ is type I this implies that $q_0 = 0$.

4.6 Proposition. *If* $M \otimes_\sigma R$ *is type I then* M *is semi-finite.*

Proof. With the notations of before let $q_1 = q - \hat{\sigma}_1(q)$. Then q_1 is a central projection with the property that $\hat{\sigma}_t(q_1)q_1 = 0$ if $|t| \ge 1$. Indeed

$$q = \hat{\sigma}_0(q) \ge \hat{\sigma}_1(q) \text{ and } \hat{\sigma}_t(q_1) = \hat{\sigma}_t(q) - \hat{\sigma}_{t+1}(q)$$

so if $t \ge 1$, $\hat{\sigma}_t(q_1) \le \hat{\sigma}_t(q) \le \hat{\sigma}_1(q)$ and $\hat{\sigma}_t(q_1)q_1 = 0$. If $t \le -1$ then $\hat{\sigma}_{t+1}(q) \ge q \ge q_1$ so that $\hat{\sigma}_t(q_1)q_1 = 0$. We also have that

$\hat{\sigma}_k(q_1) = \hat{\sigma}_k(q) - \hat{\sigma}_{k+1}(q)$ are mutually orthogonal when $k \in \mathbf{Z}$ and that $\sum_{k=-\infty}^{+\infty} \hat{\sigma}_k(q_1) = 1$ by lemma 4. 5. Now define τ_0 on M^+ by $\tau_0(x) = \tau(\pi(x)q_1)$. Because τ is a trace and q_1 is in the centre, also τ_0 will be a trace. It is also normal as τ is normal. To prove that τ_0 is faithful assume $x \in M^+$ and $\tau_0(x) = 0$, then $\tau(\pi(x)q_1) = 0$ and $\pi(x)q_1 = 0$ as τ is faithful. Also $\hat{\sigma}_k(\pi(x)q_1) = \pi(x)\hat{\sigma}_k(q_1) = 0$ for all $k \in \mathbf{Z}$ and because $\sum_{k=-\infty}^{+\infty} \hat{\sigma}_k(q_1) = 1$ it follows that $\pi(x) = 0$. Then $x = 0$ as π is faithful. (Remark that it is in fact only to prove the faithfulness of τ_0 that we need that $M \otimes_\sigma \mathbf{R}$ is type I.)

Let us now show that τ_0 is semi-finite. Therefore let f be any non-zero projection in M. If $\pi(f)q_1 = 0$ we would have as before that $\pi(f) = 0$. So by the semi-finiteness of τ there is non-zero projection $e \in M \otimes_\sigma \mathbf{R}$ such that $\tau(e) < \infty$ and $e \leq \pi(f)q_1$. Define

$$h_n = \int_{-n}^{n+1} \hat{\sigma}_t(e)dt = \int_0^1 \hat{\sigma}_t\Big(\sum_{k=-n}^{n} \hat{\sigma}_k(e)\Big)dt.$$

Because $e \leq q_1$, we also have $\hat{\sigma}_k(e) \leq \hat{\sigma}_k(q_1)$ and therefore also $\{\hat{\sigma}_k(e)\}_{k \in \mathbf{Z}}$ are mutually orthogonal. So $\sum_{-n}^{n} \hat{\sigma}_k(e)$ increases to a projection. Because of the normality of the map $\int_0^1 \hat{\sigma}_t dt$ it follows that also h_n will increase to an element $h \in M \otimes_\sigma \mathbf{R}$ with $0 \leq h \leq 1$. Obviously $\hat{\sigma}_t(h) = h$ for all $t \in \mathbf{R}$. Now $e \leq \pi(f)$ so that $\hat{\sigma}_t(e) \leq \hat{\sigma}_t(\pi(f)) = \pi(f)$. This means that $\pi(f)\hat{\sigma}_t(e)\pi(f) = \hat{\sigma}_t(e)$ and so also $\pi(f)\, h\, \pi(f) = h$, and it follows that $h \leq \pi(f)$. Now $\tau(h\, q_1) = \sup \int_{-n}^{n+1} \tau(\hat{\sigma}_t(e)q_1)dt$. But $e \leq q_1$ so that $\hat{\sigma}_t(e) \leq \hat{\sigma}_t(q_1)$ and as $\hat{\sigma}_t(q_1)q_1 = 0$ for all $|t| \geq 1$ also $\hat{\sigma}_t(e)q_1 = 0$ for all $|t| \geq 1$ and therefore

$$\begin{aligned} \tau(h\, q_1) &= \int_{-1}^{1} \tau(\hat{\sigma}_t(e)q_1)dt \\ &= \int_{-1}^{1} e^{-t}\tau(e\, \hat{\sigma}_{-t}(q_1))dt \\ &\leq \int_{-1}^{1} e^{-t}\tau(e)dt < \infty . \end{aligned}$$

Because $h \in M \otimes_\sigma \mathbf{R}$ and $\hat{\sigma}_t(h) = h$, there is an element $x \in M$ such that $\pi(x) = h$. Then $h \leq \pi(f) \Rightarrow x \leq f$ and $\tau_0(x) = \tau(h\, q_1) < \infty$. This implies that τ_0 is semi-finite on M. Then the proof is complete.

4.7 Theorem. _If_ M _is of type III, then_ $M \otimes_\sigma R$ _is of type_ II_∞.

Proof. Let $\tilde{p}$ be the largest central projection in $M \otimes_\sigma R$ such that $(M \otimes_\sigma R)\tilde{p}$ is type I. Such a projection is invariant for any $*$-automorphism and in particular $\hat{\sigma}_t(\tilde{p}) = \tilde{p}$. Then $\tilde{p}$ is of the form $\pi(p)$ with $p \in M$ and because $\pi(p)$ is central in $M \otimes_\sigma R$, we must have that p is in the centre of M. Therefore we can apply proposition 4.1 and we obtain that

$$(Mp \otimes_{\bar{\sigma}} R) = (M \otimes_\sigma R)(p \otimes 1).$$

Of course $\bar{\sigma}$ is the modular action on Mp associated with the restriction of the original functional ϕ to Mp. Now $p \otimes 1 = \pi(p)$ and $(M \otimes_\sigma R)(p \otimes 1)$ is type I and by the previous result Mp is semi-finite. Because M is type III we must have $p = 0$. So $M \otimes_\sigma R$ is type II.

Now let $\tilde{p}_1$ be the largest central projection such that $(M \otimes_\sigma R)\tilde{p}_1$ is type II_1. Again $\tilde{p}_1 = p_1 \otimes 1$ for some central $p_1 \in M$ and $Mp_1 \otimes_{\bar{\sigma}} R$ would be finite. Then Mp_1 would be finite as a subalgebra which implies $p_1 = 0$ as M is type III. This shows that $M \otimes_\sigma R$ is type II_∞.

If we combine this result with earlier results we obtain the fundamental structure theorem of Takesaki:

4.8 Theorem. _If_ M _is a type III von Neumann algebra there is a type_ II_∞ _von Neumann algebra_ M_0 _with a continuous action_ θ _of_ **R** _on_ M_0, _admitting a relatively invariant trace such that_ $M \cong M_0 \otimes_\theta R$.

Takesaki also shows that the pair (M_0, θ) is uniquely determined by M up to weak equivalence. To prove this result we would have to use the modular theory for weights. Remark however that M_0 is uniquely determined when considered as the crossed product of M with modular actions.

Appendices

Let X be a locally compact Hausdorff space, and ds denote any positive measure on X in the sense of [1]. Also let $\mathcal{H}$ be a (complex) Hilbert space.

A1 Notation. By $L_2(X, \mathcal{H})$ we denote the set of $\mathcal{H}$-valued functions ξ on X such that

(i) $\langle \xi(\cdot), \eta \rangle$ is measurable for all $\eta \in \mathcal{H}$ where $\langle \cdot , \cdot \rangle$ denotes the scalar product in $\mathcal{H}$.

(ii) there is a separable subspace $\mathcal{H}_0$ of $\mathcal{H}$ such that $\xi(s) \in \mathcal{H}_0$ for all $s \in X$.

(iii) $\int \langle \xi(s), \xi(s) \rangle ds < \infty$.

Remark that it follows from conditions (i) and (ii) that $\langle \xi(\cdot), \xi(\cdot) \rangle$ is measurable so that condition (iii) makes sense. Indeed if $\{e_n\}_{n=1,\infty}$ is an orthogonal basis in $\mathcal{H}_0$ then

$$\langle \xi(s), \xi(s) \rangle = \sum_{n=1}^{\infty} |\langle \xi(s), e_n \rangle|^2$$

and for each n we have that $\langle \xi(\cdot), e_n \rangle$ is measurable.

It is easily verified from the definition that $L_2(X, \mathcal{H})$ is a vector-space over C. In fact it can be made into a Hilbert space such that $C_c(X, \mathcal{H})$, the set of continuous $\mathcal{H}$-valued functions with compact support in X, is dense in $L_2(X, \mathcal{H})$. We will prove here those two results. (It will also justify to use the notation $L_2(G, \mathcal{H})$ in section 2 of part I.)

A2 Proposition. <u>If</u> $\xi, \eta \in L_2(X, \mathcal{H})$ <u>then</u> $\langle \xi(\cdot), \eta(\cdot) \rangle$ <u>is integrable and</u>

$$\langle \xi, \eta \rangle = \int \langle \xi(s), \eta(s) \rangle ds$$

<u>is a scalar product making</u> $L_2(X, \mathcal{H})$ <u>into a Hilbert space.</u>

We make here the usual identification of functions that are equal almost everywhere.

Proof. The measurability of $\langle \xi(\cdot), \eta(\cdot)\rangle$ follows from (i) and (ii) as above, the integrability follows from (iii) as $|\langle \xi(s), \eta(s)\rangle| \leq \|\xi(s)\| \|\eta(s)\|$. It is clear that the above formula defines a scalar product on $L_2(X, \mathcal{H})$. The only thing to show is completeness, and the proof of this is very similar to that of the completeness of $L_2(X)$.

Let ξ_n be a Cauchy sequence in $L_2(X, \mathcal{H})$. By passing to a subsequence we may assume that $\|\xi_n - \xi_{n+1}\| \leq 2^{-n}$. Define

$$g_k(s) = \sum_{n=1}^{k} \|\xi_{n+1}(s) - \xi_n(s)\|.$$

Then $g_k \in L_2(X)$ and $\|g_k\| \leq \sum_{n=1}^{k} \|\xi_{n+1} - \xi_n\| \leq 1$.

If $g(s) = \sum_{n=1}^{\infty} \|\xi_{n+1}(s) - \xi_n(s)\|$, by the monotone convergence theorem also $g \in L_2(X)$ and $\|g\| \leq 1$. In particular there is a null set E in X such that $g(s) < \infty$ for all $s \in E^c$, the complement of E. This means that $\sum_{n=1}^{\infty} (\xi_{n+1}(s) - \xi_n(s))$ converges in norm in $\mathcal{H}$ for all $s \in E^c$ or equivalently that $\lim \xi_n(s)$ exists for all $s \in E^c$.

Define $\xi(s) = \lim \xi_n(s)$ if $s \in E^c$ and $\xi(s) = 0$ if $s \in E$, then ξ will be the limit of ξ_n in $L_2(X, \mathcal{H})$. Indeed for all $\eta \in \mathcal{H}$ we have that $\langle \xi(s), \eta\rangle = \lim\langle \xi_n(s), \eta\rangle(1 - \chi_E(s))$ where χ_E is the characteristic function of E. Hence $\langle \xi(\cdot), \eta\rangle$ is measurable. For each n there is a separable subspace $\mathcal{H}_n$ of $\mathcal{H}$ such that $\xi_n(s) \in \mathcal{H}_n$ for all s. If now $\mathcal{H}_0$ is the smallest subspace containing all the $\mathcal{H}_n$ then $\mathcal{H}_0$ is still separable and $\xi(s) \in \mathcal{H}_0$ for all $s \in X$. Finally, using Fatou's lemma

$$\int \|\xi(s) - \xi_n(s)\|^2 ds \leq \liminf_m \int \|\xi_m(s) - \xi_n(s)\|^2 ds = \liminf_m \|\xi_m - \xi_n\|^2$$

so that given $\varepsilon > 0$ there is a n_0 such that if $n > n_0$ we have

$$\int \|\xi(s) - \xi_n(s)\|^2 ds < \varepsilon .$$

This implies that $\xi - \xi_n \in L_2(X, \mathcal{H})$, hence $\xi \in L_2(X, \mathcal{H})$ and also that

$\|\xi - \xi_n\| \to 0.$

Next we show that the set $C_c(X, \mathcal{H})$ of continuous $\mathcal{H}$-valued functions with compact support in X is dense in $L_2(X, \mathcal{H})$.

A3 **Proposition.** $C_c(X, \mathcal{H})$ <u>is dense in</u> $L_2(X, \mathcal{H})$.

Proof. Let us first show that $C_c(X, \mathcal{H})$ is contained in $L_2(X, \mathcal{H})$. Therefore let $\xi \in C_c(X, \mathcal{H})$, then the range of ξ is contained in a compact subset $\mathcal{K}$ of $\mathcal{H}$. Now such a compact set $\mathcal{K}$ must lie in a separable subspace of $\mathcal{H}$. To see this choose for every n a finite number of vectors $\xi_1^{(n)}, \xi_2^{(n)}, \ldots, \xi_{k_n}^{(n)}$ in $\mathcal{H}$ such that balls with radius $\frac{1}{n}$ centred at those points cover $\mathcal{K}$. Then clearly $\mathcal{K}$ is a subset of the subspace $\mathcal{H}_0$ generated by $\{\xi_j^{(n)} | j = 1, k_n; n = 1, \infty\}$. Hence we have shown that the range of ξ lies in a separable subspace. The other two conditions follow immediately so that $\xi \in L_2(X, \mathcal{H})$.

To prove the density of $C_c(X, \mathcal{H})$, assume $\xi \in L_2(X, \mathcal{H})$ and that $\langle \xi, \eta \rangle = 0$ for all $\eta \in C_c(X, \mathcal{H})$. Now let $\xi_0 \in \mathcal{H}$ and $f \in C_c(X)$ and let η be defined by $\eta(s) = f(s)\xi_0$. So $\langle \xi, \eta \rangle = \int \overline{f(s)} \langle \xi(s), \xi_0 \rangle ds = 0$. This is true for all $f \in C_c(X)$ and so $\langle \xi(s), \xi_0 \rangle = 0$ a. e.. Choose an orthonormal basis $\{e_n\}_{n=1, \infty}$ in the separable subspace containing the range of ξ. So for any n there is a null set E_n such that $\langle \xi(s), e_n \rangle = 0$ for all $s \in E_n^c$. Put $E = \bigcup_{n=1}^{\infty} E_n$, then $\langle \xi(s), e_n \rangle = 0$ for all $s \in E^c$ and all n. Hence $\xi(s) = 0$ for all $s \in E^c$ so that $\xi = 0$ as a vector in $L_2(X, \mathcal{H})$.

This proposition justifies the introduction of $L_2(X, \mathcal{H})$ as the completion of $C_c(X, \mathcal{H})$.

APPENDIX B

Let G be a locally compact group and let ds denote a left invariant Haar measure on G. Denote by N the von Neumann algebra consisting of all multiplications on $L_2(G)$ by $L_\infty(G)$-functions. Then $N = N'$, see for example [6].

B1 Lemma. $N = \{m_f | f \in C_c(G)\}''$, <u>where</u> m_f <u>is multiplication by</u> f <u>on</u> L_2.

Proof. Let ϕ be a σ-weakly continuous linear functional on N such that $\phi(m_f) = 0$ for all $f \in C_c(G)$. Any such ϕ is of the form $\phi = \sum_{n=1}^{\infty} \langle \cdot f_n, g_n \rangle$ with $f_n, g_n \in L_2(G)$, and $\sum_{n=1}^{\infty} \|f_n\|^2 < \infty$ and $\sum_{n=1}^{\infty} \|g_n\|^2 < \infty$. It follows that $h(s) = \sum_{n=1}^{\infty} f_n(s)\overline{g_n(s)}$ defines an L_1-function and we have for any $f \in C_c(G)$ that

$$\phi(m_f) = \sum_{n=1}^{\infty} \int f(s)f_n(s)\overline{g_n(s)}ds = \int f(s)h(s)ds = 0.$$

This implies that $h = 0$ as an element in L_1. Then for any $f \in L_\infty$ the above relation implies also that $\phi(m_f) = 0$ and hence $\phi = 0$.

It follows immediately that $\mathcal{B}(L_2(G)) = \{m_f\lambda_s | f \in C_c(G),\ s \in G\}''$ because if x commutes with all m_f, $f \in C_c(G)$ it must be an element in N and if this also commutes with all λ_s with $s \in G$ it must be a multiple of the identity. This result was used in section 3 of part I.

Now let G be commutative.

B2 Lemma. $N = \{v_p | p \in \hat{G}\}''$.

Proof. As in the previous lemma let ϕ be a σ-weakly continuous linear functional on N such that $\phi(v_p) = 0$ for all $p \in \hat{G}$. With the same notations as above we get $\phi(v_p) = \int \overline{\langle s, p \rangle} h(s)ds = 0$. Now the Fourier transform is injective on L_1 so that $h = 0$, and again $\phi = 0$.

This result was used in section 4 of part I.

APPENDIX C

If M is a properly infinite von Neumann algebra, then there is a sequence $\{e_n\}_{n=1,\infty}$ of mutually orthogonal projections in M such that $\sum_{n=1}^{\infty} e_n = 1$ and $e_n \sim 1$ in the sense of equivalence of projections ([6, p. 298]). Then we have the following.

Theorem. _If_ $\mathcal{H}$ _is a separable Hilbert space and_ F_∞ _denotes_ $\mathcal{B}(\mathcal{H})$, _then_ $M \otimes F_\infty \cong M$.

Proof. Let e_n be as above and let v_n be elements in M such that $e_n = v_n^* v_n$ and $v_n v_n^* = 1$. If $n \neq m$ then $e_n e_m = 0$ so that $v_m v_n^* = v_m e_m e_n v_n^* = 0$. Let $\{e_{ij}\}_{i,j=1,\infty}$ be matrix units in F_∞. Then define

$$u_k = \sum_{i=1}^{k} v_i \otimes e_{i1}.$$

Then

$$u_k^* u_k = (\sum_{i=1}^{k} v_i^* \otimes e_{1i})(\sum_{j=1}^{k} v_j \otimes e_{j1}) = \sum_{i,j=1}^{k} v_i^* v_j \otimes e_{1i} e_{j1}$$

$$= \sum_{i=1}^{k} v_i^* v_i \otimes e_{11} = (\sum_{i=1}^{k} e_i) \otimes e_{11}.$$

Similarly

$$u_k u_k^* = (\sum_{i=1}^{k} v_i \otimes e_{i1})(\sum_{j=1}^{k} v_j^* \otimes e_{1j})$$

$$= \sum_{i,j=1}^{k} v_i v_j^* \otimes e_{ij} = \sum_{i=1}^{k} v_i v_i^* \otimes e_{ii}$$

$$= 1 \otimes \sum_{i=1}^{k} e_{ii}.$$

A similar calculation would show that, if $k > l$,

$$(u_k^* - u_l^*)(u_k - u_l) = \sum_{i=l+1}^{k} e_i \otimes e_{11}$$

$$(u_k - u_l)(u_k^* - u_l^*) = 1 \otimes \sum_{i=l+1}^{k} e_{ii}.$$

Because both $\sum_{i=1}^{\infty} e_i$ and $\sum_{i=1}^{\infty} e_{ii}$ converge we obtain that $\{u_k\}$ is a Cauchy sequence in the strong*-topology. Hence if u is the limit, then $u = \lim u_k$ and $u^* = \lim u_k^*$.

It will then follow that u is an element in $M \otimes F_\infty$ such that $u^* u = 1 \otimes e_{11}$ and $uu^* = 1 \otimes 1$.

Now define $\sigma(\tilde{x}) = u^* \tilde{x} u$ for $\tilde{x} \in M \otimes F_\infty$. Then as $u^* u = 1 \otimes e_{11}$

the range of σ is

$$(1 \otimes e_{11})(M \otimes F_{\infty})(1 \otimes e_{11}) = M \otimes e_{11}.$$

As $uu^* = 1$, σ will be a homomorphism of $M \otimes F_{\infty}$ onto $M \otimes e_{11}$. If $\sigma(\tilde{x}) = 0$, $u^*\tilde{x}\,u = 0$, then $\tilde{x} = uu^*\,\tilde{x}\,uu^* = 0$. So σ is an isomorphism. Now as $M \otimes e_{11} \simeq M$ we have the result.

APPENDIX D

Let M be a semi-finite von Neumann algebra, ϕ a faithful normal positive linear functional on M and σ the associated modular action. We will show in this appendix that σ is inner.

Let τ be a faithful normal semi-finite trace on M and let e be a projection in M such that $\tau(e) < \infty$. Then the restriction of τ to $e\,M\,e$ is a faithful normal finite trace on $e\,M\,e$.

D1 **Lemma.** _If_ τ _is a faithful normal finite trace on_ M _and_ ϕ _a faithful normal positive linear functional on_ M, _there is a unique_ $h \in M$ _such that_ $0 \leq h \leq 1$ _and_ $\tau((1 - h)x) = \phi(h\,x) = \phi(x\,h)$ _for all_ $x \in M$.

Proof. Apply Sakai's linear Radon Nikodym theorem [14] to the functionals τ and $\phi + \tau$. So there is an $h \in M$ such that $0 \leq h \leq 1$ and

$$\tau(x) = \tfrac{1}{2}(\phi + \tau)(xh + hx)$$

$$= \tau(xh) + \tfrac{1}{2}\phi(xh + hx).$$

Then $\tau((1 - h)x) = \tfrac{1}{2}\phi(xh + hx)$. Replacing first x by hx and next x by xh we get

$$\tau((1 - h)hx) = \tfrac{1}{2}\phi(hxh + h^2x)$$

and

$$\tau((1 - h)xh) = \tfrac{1}{2}\phi(xh^2 + h\,x\,h).$$

As $\tau((1 - h)hx) = \tau((1 - h)xh)$ we obtain $\phi(xh^2) = \phi(h^2x)$ for all $x \in M$. From this it follows that also $\phi(xh) = \phi(hx)$ for all x. Indeed we would get $\phi(xh^4) = \phi(h^2xh^2) = \phi(h^4x)$, and similarly for any even power $\phi(xh^{2n}) = \phi(h^{2n}x)$ and approximating h by polynomials in h^2 we get the desired result.

So we get $\tau((1 - h)x) = \phi(hx) = \phi(xh)$ for all $x \in M$. To prove uniqueness, suppose also $\tau((1 - k)x) = \phi(kx)$ for all $x \in M$ with $k \in M$. Then $\tau(x) = (\tau + \phi)(kx) = (\tau + \phi)(hx)$ for all x. As $\tau + \phi$ is faithful this implies $k = h$.

D2 Lemma. Let τ be a faithful normal semi-finite trace on M and ϕ a faithful normal positive linear functional on M. Then there is a $h \in M$ with $0 \le h \le 1$ such that $\tau(1 - h) < \infty$ and $\phi(hx)=\phi(xh)=\tau((1-h)x)$. Moreover h and $1 - h$ are injective.

Proof. Choose a net $\{e_\alpha\}$ of projections in M such that $\tau(e_\alpha) < \infty$ and e_α increases up to 1. Then apply the previous lemma to the restrictions of ϕ and τ to $e_\alpha M e_\alpha$ to obtain an element $h_\alpha \in e_\alpha M e_\alpha$ with $0 \le h_\alpha \le 1$ such that $\tau((e_\alpha - h_\alpha)e_\alpha x e_\alpha) = \phi(h_\alpha e_\alpha x e_\alpha)$ for all $x \in M$. That is the same as

$$\tau((1 - h_\alpha)x\, e_\alpha) = \phi(h_\alpha\, x\, e_\alpha)$$

for all $x \in M$.

Because the unit ball in M is σ-weakly compact we can assume (if necessary by taking a subnet) that h_α converges σ-weakly to an element $h \in M$ with $0 \le h \le 1$.

Now if $e_\beta \le e_\alpha$ then $xe_\beta = x\, e_\beta\, e_\alpha$ and we get

$$\tau((1 - h_\alpha)x\, e_\beta) = \tau((1 - h_\alpha)x\, e_\beta e_\alpha) = \phi(h_\alpha\, x\, e_\beta e_\alpha) = \phi(h_\alpha\, x\, e_\beta).$$

Because $\tau(e_\beta) < \infty$ we have that $\tau(\cdot\, e_\beta)$ is σ-weakly continuous so that if we take the limit over α we find $\tau((1 - h)x\, e_\beta) = \phi(h\, x\, e_\beta)$ and with $x = 1$ in particular we get $\tau((1 - h)e_\beta) = \phi(h\, e_\beta)$. By normality $\tau((1 - h)e_\beta)$ will increase to $\tau(1 - h)$ and therefore $\tau(1 - h) = \phi(h) < \infty$. Therefore also $\tau((1 - h)\cdot)$ is σ-weakly continuous and in the limit over β we get

$$\tau((1-h)x) = \phi(hx) \quad \text{for all} \quad x \in M.$$

Similarly or by taking adjoints we get $\tau((1-h)x) = \phi(xh)$.

Finally h and $1-h$ must be injective as ϕ and τ are faithful.

Of course we have essentially proved here the well known Radon-Nikodym theorem for semi-finite von Neumann algebras.

D3 Lemma. *With τ, ϕ and h as in lemma 2, the modular automorphism group associated to ϕ is given by*

$$\sigma_t(x) = h^{-it}(1-h)^{it} \, x \, h^{it}(1-h)^{-it}$$

for all $x \in M$ and $t \in R$.

Proof. By spectral theory one can show that h^{iz} is well defined for $\operatorname{Im} z \leq 0$, that it is continuous and uniformly bounded on finite horizontal strips, and that it is analytic for $\operatorname{Im} z < 0$, with respect to the strong topology [13]. Similarly for $(1-h)^{iz}$. Then it follows that for any $x \in M$ the function $z \to h^{-iz}(1-h)^{iz+1} \, x \, h^{iz+1}(1-h)^{-iz}$ is well defined, bounded and strongly continuous for $\operatorname{Im} z \in [0, 1]$, and strongly analytic inside this strip. Then for any pair $x, y \in M$ the function F defined by

$$F(z) = \phi(h^{-iz}(1-h)^{iz+1} x \, h^{iz+1}(1-h)^{-iz} y)$$

is well-defined, bounded and continuous for $\operatorname{Im} z \in [0, 1]$, and analytic inside. Moreover, if we define σ_t by the above expression, we find

$$\begin{aligned} F(t) &= \phi(h^{-it}(1-h)^{it}(1-h)x \, h^{it} h(1-h)^{-it} y) \\ &= \phi(\sigma_t((1-h)x \, h)y) \end{aligned}$$

and

$$\begin{aligned} F(t+i) &= \phi(h^{-it} h(1-h)^{it} \, x \, h^{it}(1-h)^{-it}(1-h)y) \\ &= \phi(h \, \sigma_t(x)(1-h)y) \\ &= \tau((1-h)\sigma_t(x)(1-h)y) \\ &= \tau((1-h)y(1-h)\sigma_t(x)) \end{aligned}$$

$$= \phi(h\, y(1-h)\sigma_t(x))$$
$$= \phi(y(1-h)\sigma_t(x)h)$$
$$= \phi(y\, \sigma_t((1-h)x\, h)).$$

Therefore the K. M. S. -condition is satisfied for any pair of the form $((1-h)xh,\ y)$. We will now show by an approximation argument that the K. M. S. -condition is also satisfied for any pair (x, y), see also [13].

We first need the σ_t-invariance of ϕ. This can be obtained directly from the definitions. It also follows from the K. M. S. -condition applied to the pair $((1-h)xh,\ 1)$. Indeed as in the remark following theorem 2. 1 of part II we obtain that $\phi(\sigma_t((1-h)xh)) = \phi((1-h)xh)$ for all $t \in \mathbf{R}$. Now because $1-h$ and h are injective any $x \in M$ can be approximated strongly by a bounded sequence $\{x_n\}$ in $(1-h)Mh$ and therefore also $\phi(\sigma_t(x)) = \phi(x)$ for all $t \in \mathbf{R}$ and $x \in M$.

To obtain the K. M. S. -function for the pair (x, y) in M, consider the K. M. S. -functions F_n associated to the pairs (x_n, y) where again $\{x_n\}$ is a bounded sequence in $(1-h)Mh$ converging strongly to x.

Then, using the invariance of ϕ we get

$$|F_n(t) - F_m(t)| = |\phi(\sigma_t(x_n - x_m)y))| \le \phi((x_n - x_m)^*(x_n - x_m))^{\frac{1}{2}}\phi(y^*y)^{\frac{1}{2}}$$

and $F_n(t) - F_m(t) \to 0$ uniformly in t. Similarly $F_n(t+i) - F_m(t+i) \to 0$ uniformly in t and therefore by the maximum modulus principle for the strip it follows that $F_n(z) - F_m(z) \to 0$ uniformly in z when $n, m \to \infty$. Then $F(z) = \lim F_n(z)$ will define the right K. M. S. -function for the pair (x, y). Finally because the modular automorphism group is the unique strongly continuous one parameter group of *-automorphisms satisfying the K. M. S. -condition with respect to ϕ the result follows.

D4 Theorem. <u>If</u> M <u>is semi-finite, then every modular action is inner.</u>

Proof. Follows immediately from lemma D3.

References

[1] N. Bourbaki, Intégration, Ch. 1-4, Paris (1952) and Intégration, Ch. 7-8, Paris (1963).

[2] A. Connes, Une classification des facteurs de type III. Ann. Sc. de l'Ecole Normal Supérieur, 4 série, t. 6 (1973), 133-252.

[3] A. Connes, A factor not anti-isomorphic to itself. The Bull. of the London Math. Soc., 7 (1975), 171-4.

[4] T. Digernes, Poids dual sur un produit croisé. C. R. Acad. Sc. Paris, t. 278A (1974), 937-40.

[5] T. Digernes, Duality for weights on covariant systems and its applications (1975), UCLA Thesis.

[6] J. Dixmier, Les algèbres d'opérateurs dans l'espace Hilbertien, 2e ed. Paris, Gauthier Villars (1969).

[7] S. Doplicher, D. Kastler, D. Robinson, Covariance algebras in field theory and statistical mechanics. Comm. Math. Phys. 3 (1966) 1-28.

[8] U. Haagerup, On the dual weights for crossed products of von Neumann algebras I, II, (1975). Preprints Odense University, Denmark.

[9] P. Halmos, Measure theory, Van Nostrand, London (1950).

[10] J. Kelley, General topology, Van Nostrand Reinhold, London (1955).

[11] L. Loomis, An introduction to abstract harmonic analysis. Van Nostrand, London (1953).

[12] M. Rieffel, A. Van Daele, The commutation theorem for tensor products of von Neumann algebras, Bull. London Math. Soc. 7 (1975), 257-60.

[13] M. Rieffel, A. Van Daele, A bounded operator approach to the Tomita-Takesaki theory. Pac. Journal Math. 69 (1977), 187-221.

[14] S. Sakai, C*- and W*-algebras. Ergebnise der Mathematik und ihrer Grenzgebite, Band 60, Springer-Verlag, Berlin (1971).

[15] M. Takesaki, Tomita's theory of modular Hilbert algebras and its applications, Lecture Notes in Mathematics 128 (1970), Springer-Verlag, Berlin.

[16] M. Takesaki, Duality for crossed products and the structure of von Neumann algebras of type III. Acta Mathematica 131 (1973), 249-310.

[17] A. Van Daele, The Tomita-Takesaki theory for von Neumann algebras with a separating and cyclic vector. Proceedings of the international school of physics Enrico Fermi, Course LX, C*-algebras and their applications to statistical mechanics and quantum field theory. North-Holland, Amsterdam (1976).

[18] A. Van Daele, Fixed points and commutation theorems, Second Japan U. S. seminar on C*-algebras and applications to physics, Los Angeles (1977).

Index

INDEX OF NOTATIONS